T0214343

Lecture Notes in Computer Science 11292

Commenced Publication in 1973
Founding and Former Series Editors:
Gerhard Goos, Juris Hartmanis, and Jan van Leeuwen

More information about this series at http://www.springer.com/series/7409

Yuen-Hsien Tseng · Tetsuya Sakai
Jing Jiang · Lun-Wei Ku
Dae Hoon Park · Jui-Feng Yeh
Liang-Chih Yu · Lung-Hao Lee
Zhi-Hong Chen (Eds.)

Information Retrieval Technology

14th Asia Information Retrieval Societies Conference, AIRS 2018
Taipei, Taiwan, November 28–30, 2018
Proceedings

 Springer

Editors
Yuen-Hsien Tseng
National Taiwan Normal University
Taipei, Taiwan

Tetsuya Sakai
Waseda University
Tokyo, Japan

Jing Jiang
School of Information Systems
Singapore Management University
Singapore, Singapore

Lun-Wei Ku
Academia Sinica
Taipei, Taiwan

Dae Hoon Park
Huawei Research America
Champaign, IL, USA

Jui-Feng Yeh
National Chiayi University
Chiayi City, Taiwan

Liang-Chih Yu
Yuan Ze University
Taoyuan City, Taiwan

Lung-Hao Lee
National Central University
Taoyuan City, Taiwan

Zhi-Hong Chen
National Taiwan Normal University
Taipei City, Taiwan

ISSN 0302-9743 ISSN 1611-3349 (electronic)
Lecture Notes in Computer Science
ISBN 978-3-030-03519-8 ISBN 978-3-030-03520-4 (eBook)
https://doi.org/10.1007/978-3-030-03520-4

Library of Congress Control Number: 2018960429

LNCS Sublibrary: SL3 – Information Systems and Applications, incl. Internet/Web, and HCI

This Springer imprint is published by the registered company Springer Nature Switzerland AG
The registered company address is: Gewerbestrasse 11, 6330 Cham, Switzerland

Preface

Welcome to the proceedings of the 14th Asia Information Retrieval Societies conference (AIRS 2018)! AIRS is the premier Asian regional conference in the broad area of information retrieval (IR). Following the successes of IROL 1996 (International Workshop on IR with Oriental Languages), and IRAL (International Workshop on IR with Asian Languages, 1997–2003), the first AIRS (Asia IR *Symposium*) was held in 2004 in Beijing. In 2010, the first Asia Information Retrieval Societies Conference (AIRS 2010) was held in Taipei. And after eight years, we returned to Taipei!

The acceptance rates for AIRS 2018 were 34% (8/24) for full papers and 53% (9/17) for short/demo papers. Each paper was reviewed by three Program Committee (PC) members and the final decisions were made by the PC chairs (for full papers) and the short-paper PC chairs (for short and demo papers). Given the limited number of accepted papers, this year we decided to have more flexible page limits for camera-ready papers compared with previous years: 12 pages plus references for full papers, and 6 pages plus references for short/demo papers. The best paper (for full papers only) was selected by the PC chairs and the short-paper PC chairs by considering the review scores.

The AIRS 2018 proceedings also includes three short papers from AIRS 2017 that were inadvertently excluded from the AIRS 2017 proceedings published last year.

We would like to thank our PC members, general chair, publication chair, local chairs, and the AIRS Steering Committee for helping us compile an exciting conference program. Enjoy!

November 2017

Yuen-Hsien Tseng
Tetsuya Sakai
Jing Jiang
Lun-Wei Ku
Dae Hoon Park
Jui-Feng Yeh
Liang-Chih Yu
Lung-Hao Lee
Zhi-Hong Chen

Organization

General Chair

Yuen-Hsien Tseng   National Taiwan Normal University, Taiwan

Program Co-chairs

Tetsuya Sakai   Waseda University, Japan
Jing Jiang   Singapore Management University, Singapore

Poster and Demo Co-chairs

Lun-Wei Ku   Academia Sinica, Taiwan
Dae Hoon Park   Huawei Research America, USA

Publication Chair

Jui-Feng Yeh   National Chiayi University, Taiwan

Local Co-chairs

Liang-Chih Yu   Yuan Ze University, Taiwan
Lung-Hao Lee   National Central University, Taiwan
Zhi-Hong Chen   National Taiwan Normal University, Taiwan

AIRS 2018 Program Committee

Hiroyuki Toda   NTT, Japan
Jaewon Kim   RMIT University, Australia
Kazunari Sugiyama   National University of Singapore, Singapore
Harumi Murakami   Osaka City University, Japan
Atsushi Keyaki   Tokyo Institute of Technology, Japan
Suthee Chaidaroon   Santa Clara University, USA
Yue Wang   University of Michigan, USA
Xiangnan He   National University of Singapore, Singapore
Yohei Seki   University of Tsukuba, Japan
Makoto P. Kato   Kyoto University, Japan
Cheng Luo   Tsinghua University, China
Deguang Kong   Yahoo Research, USA
Takehiro Yamamoto   Kyoto University, Japan
Liu Yang   University of Massachusetts Amherst, USA
Zhicheng Dou   Renmin University of China, China

Chenliang Li	Wuhan University, China
Koji Eguchi	Kobe University, Japan
Hyun Joon Jung	Apple, USA
Hiroaki Ohshima	University of Hyogo, Japan
Hidetsugu Nanba	Hiroshima City University, Japan
Mengwen Liu	Drexel University, USA
Yiqun Liu	Tsinghua University, China
Yi Fang	Santa Clara University, USA
Masayuki Okamoto	Corporate R&D Center, Toshiba Corporation
Sheng Wang	Peking University, China
Pu-Jen Cheng	National Taiwan University, Taiwan
Jian-Yun Nie	University of Montreal, Canada
Li Su	Academia Sinica, Taiwan
Hideo Joho	University of Tsukuba, Japan
Adam Jatowt	Kyoto University, Japan
Chien Chin Chen	National Taiwan University, Taiwan
Sumio Fujita	Yahoo! JAPAN Research, Japan
Tadashi Nomoto	National Institute of Japanese Literature, Japan
Aston Zhang	Amazon AI, USA
Ben He	University of Chinese Academy of Sciences, China
Xianpei Han	Institute of Software, Chinese Academy of Sciences, China
Hui Fang	University of Delaware, USA
Masaharu Yoshioka	Hokkaido University, Japan
Grace Hui Yang	Georgetown University, USA
Xin Zhao	Renmin University of China, China
Min Zhang	Tsinghua University, China
Qingyao Ai	University of Massachusetts Amherst, USA
Jun Harashima	Cookpad Inc., Japan
Aixin Sun	Nanyang Technological University, Singapore
Jaap Kamps	University of Amsterdam, The Netherlands
Nancy-I-Chin Wu	National Taiwan Normal University, Taiwan
Andrew Trotman	University of Otago, New Zealand
Xuanjing Huang	Fudan University, China
Udo Kruschwitz	University of Essex, UK
Mark Sanderson	RMIT University, Australia

AIRS Steering Committee

Hsin-Hsi Chen	National Taiwan University, Taiwan
Zhicheng Dou	Renmin University of China, China
Wai Lam	The Chinese University of Hong Kong, SAR China
Alistair Moffat	University of Melbourne, Australia
Hwee Tou Ng	National University of Singapore, Singapore
Dawei Song	The Open University, UK
Masaharu Yoshioka	Hokkaido University, Japan

Contents

Social

An Ensemble Neural Network Model for Benefiting Pregnancy Health
Stats from Mining Social Media 3
  Neha Warikoo, Yung-Chun Chang, Hong-Jie Dai, and Wen-Lian Hsu

Chinese Governmental Named Entity Recognition 16
  Qi Liu, Dong Wang, Meilin Zhou, Peng Li, Baoyuan Qi, and Bin Wang

StanceComp: Aggregating Stances from Multiple Sources
for Rumor Detection 29
  Hao Xu and Hui Fang

Personalized Social Search Based on Agglomerative Hierarchical
Graph Clustering 36
  Kenkichi Ishizuka

Search

Improving Session Search Performance with a Multi-MDP Model 45
  *Jia Chen, Yiqun Liu, Cheng Luo, Jiaxin Mao, Min Zhang,
  and Shaoping Ma*

Analysis of Relevant Text Fragments for Different Search Task Types 60
  Atsushi Keyaki and Jun Miyazaki

A FAQ Search Training Method Based on Automatically
Generated Questions 67
  *Takuya Makino, Tomoya Noro, Hiyori Yoshikawa, Tomoya Iwakura,
  Satoshi Sekine, and Kentaro Inui*

Embedding

A Neural Labeled Network Embedding Approach to Product
Adopter Prediction 77
  Qi Gu, Ting Bai, Wayne Xin Zhao, and Ji-Rong Wen

RI-Match: Integrating Both Representations and Interactions
for Deep Semantic Matching 90
  *Lijuan Chen, Yanyan Lan, Liang Pang, Jiafeng Guo, Jun Xu,
  and Xueqi Cheng*

Modeling Relations Between Profiles and Texts   103
    Minoru Yoshida, Kazuyuki Matsumoto, and Kenji Kita

Recommendation and Classification

Missing Data Modeling with User Activity and Item
Popularity in Recommendation   113
    Chong Chen, Min Zhang, Yiqun Liu, and Shaoping Ma

Influence of Data-Derived Individualities on Persuasive Recommendation . . .   126
    Masashi Inoue and Hiroshi Ueno

Guiding Approximate Text Classification Rules via Context Information   133
    Wai Chung Wong, Sunny Lai, Wai Lam, and Kwong Sak Leung

Medical and Multimedia

Key Terms Guided Expansion for Verbose Queries in Medical Domain   143
    Yue Wang and Hui Fang

Ad-hoc Video Search Improved by the Word Sense Filtering
of Query Terms   157
    Koji Hirakawa, Kotaro Kikuchi, Kazuya Ueki, Tetsunori Kobayashi,
    and Yoshihiko Hayashi

Considering Conversation Scenes in Movie Summarization   164
    Masashi Inoue and Ryu Yasuhara

Best Paper Session

Hierarchical Attention Network for Context-Aware Query Suggestion   173
    Xiangsheng Li, Yiqun Liu, Xin Li, Cheng Luo, Jian-Yun Nie, Min Zhang,
    and Shaoping Ma

Short Papers from AIRS 2017

Assigning NDLSH Headings to People on the Web   189
    Masayuki Shimokura and Harumi Murakami

MKDS: A Medical Knowledge Discovery System Learned from Electronic
Medical Records (Demonstration)   196
    Hen-Hsen Huang, An-Zi Yen, and Hsin-Hsi Chen

Predicting Next Visited Country of Twitter Users   203
    Muhammad Syafiq Mohd Pozi, Yuanyuan Wang, Panote Siriaraya,
    Yukiko Kawai, and Adam Jatowt

Author Index   211

Social

An Ensemble Neural Network Model for Benefiting Pregnancy Health Stats from Mining Social Media

Neha Warikoo[1,2,3], Yung-Chun Chang[4(✉)], Hong-Jie Dai[5], and Wen-Lian Hsu[3]

[1] Institute of Biomedical Informatics, National Yang-Ming University,
Taipei 112, Taiwan
nehawarikoo@iis.sinica.edu.tw
[2] Bioinformatics Program, International Graduate Program Taiwan,
Institute of Information Science, Academia Sinica, Taipei 115, Taiwan
[3] Institute of Information Science, Academia Sinica, Taipei 115, Taiwan
{nehawarikoo,hsu}@iis.sinica.edu.tw
[4] Graduate Institute of Data Science, Taipei Medical University,
Taipei 106, Taiwan
changyc@tmu.edu.tw
[5] Department of Computer Science and Information Engineering,
National Taitung University, Taitung, Taiwan
hjdai@nttu.edu.tw

Abstract. Extensive use of social media for communication has made it a desired resource in human behavior intensive tasks like product popularity, public polls and more recently for public health surveillance tasks such as lifestyle associated diseases and mental health. In this paper, we exploited Twitter data for detecting pregnancy cases and used tweets about pregnancy to study trigger terms associated with maternal physical and mental health. Such systems can enable clinicians to offer a more comprehensive health care in real time. Using a Twitter-based corpus, we have developed an ensemble Long-short Term Memory (LSTM) – Recurrent Neural Networks (RNN) and Convolution Neural Networks (CNN) network representation model to learn legitimate pregnancy cases discussed online. These ensemble representations were learned by a SVM classifier, which can achieve F_1-score of 95% in predicting pregnancy accounts discussed in tweets. We also further investigate the words most commonly associated with physical disease symptoms 'Distress' and negative emotions 'Annoyed' sentiment. Results from our sentiment analysis study are quite encouraging, identifying more accurate triggers for pregnancy sentiment classes.

Keywords: Ensemble deep learning · Text mining of Twitter data
Sentiment analysis · Health surveillance · Pregnancy health stats

© Springer Nature Switzerland AG 2018
Y.-H. Tseng et al. (Eds.): AIRS 2018, LNCS 11292, pp. 3–15, 2018.
https://doi.org/10.1007/978-3-030-03520-4_1

1 Introduction

The exchange of information on an array of social media platforms like Facebook, Twitter, Youtube, Instagram has brought forth a new way of human communication; following each other's digital footprints. Between them, Twitter[1] and Facebook[2] cater about 2.2 billion users per month, out of which about 30% belong to an age demographic of 30 years or less. People use such forums to discuss all manner of topics including social issues like human rights, maternal health, child mortality etc. For example, a twitter data study in 2016 revealed maternal health issue as one of the prominent concerns with 110376 tweets discussing the issue[3]. The web-based resources capturing data through non-traditional channels of news reports and published case stories have been employed in past to disseminate health alerts to people. Although news media supplemented health information is crucial in digital age, big data institutions like Google and Facebook have made search engines new trend in health services with about 37-52% of people relying on search engines for health information on continental U.S [3].

In addition, interactive online forums have made the online peer consultation option more appealing. One study indicated that about 26% of the adults choose social media platforms for discussing health information and women contribute about 90% of that demographic[4]. Their study also revealed that 60% of health information discussion by women on such forums is related to pregnancy support or apps. Social media mining for medical conditions has become quite popular, making medical web surveillance an increasingly effective way of health data collection. Where traditional methods are limited to annual or bi-annual collection, data collection from such sources is consistent and often helpful in tracking changes in medical health status.

In a world with 130 million births reported every year, women's health and medical conditions associated with pregnancies should be studied from every possible source. Nowadays where women have also become an avid consumer of social media platforms, it seems logical to leverage such resources for improvising health statistics on pregnancies. Significant health stats particular to trimesters can help medical professionals to offer better health care and mitigate medical conditions preemptively. Keeping up with the popular themes in marketing, retailers can use such data to offer more customized shopping products for expecting mothers. There are multiple aspects to pregnancy term, which can be understood via alternative platforms like social media where people willingly and frequently discuss various issues like medical conditions to behavioral changes, support groups etc. The potential of social media to impact maternal health care system in this instance presents a huge motivation for developing an effective learning model to identify pregnancy related cases and nature of their concerns. Below are our goals for this study:

---

[1] https://zephoria.com/twitter-statistics-top-ten/.

[2] https://newsroom.fb.com/company-info/.

[3] https://www.soas.ac.uk/blogs/study/twitter-study-un-real-world-issues/.

[4] http://www.businesswire.com/news/home/20121120005872/en/Twenty-percent-online-adults discuss-health-information, 2012.

- We intend to develop a robust learning model that can detect the relevance of a discourse to legitimate pregnancy related discussion or concern.
- We would like to identify concerns among pregnant women be it physical or mental health, and annotate them as sentiments. Afterwards we can identify the key terms or causes associated with such sentiments.

2 Related Work

Digital surveillance has been around since the early days of Internet when news items and local reports were used to disseminate health information via software tools like ProMED-mail and GPHIN [1–3]. Recently, owing to the increased use of social media-based monitoring systems, international health organizations have been able to detect and monitor the progress of potential epidemics like dengue fever [5], *E.Coli* [6] and Ebola [7] outbreaks. Social media mining is singularly productive in cases where healthcare and government officials don't have sufficient outbreak data or don't have enough resources. With the outbreak of Zika in Latin America, subsequent reports of Zika cases in U.S and unavailability of governments stats on Zika, unofficial data from social media was instrumental in modeling independent Zika surveillance [8]. Social media mining has proved to be dexterous in identifying medically significant trends in lifestyle or behavior related diseases. A study by Mejova *et al.* [9] showed how culinary choices are associated with social interactions and perceptions particularly in obesity cases. Similarly, studies have been conducted on trends in alcohol abuse [10]. For adverse drug reaction (ADR) detection using social media, some systems identify dedicated drug-based triggers and establish lexicons to determine the true meaning behind the symptoms and drug association [11]. In some instances, the sizable data offered by social media sites also encourages use of supervised learning methods like Naïve Bayes and SVM [12–14].

Large-scale studies associated with pregnancy are relatively sporadic with primary means of data collections limited to traditional surveillance. Banjari *et al.* [15] identified age and pre-pregnancy BMI as compounding factors for maternity term safety using hierarchical clustering of first trimester clinical samples from pregnant women. In another study, Laopaiboon *et al.* [16] used categorized multivariate logistic regression to identify high-risk pregnancies based on age and clinical health records. Another study based on patient health records identified pregnancies associated with overall 202 fetal disorders and ADR [17]. The traditional data from clinical records used in such studies is difficult and expensive to obtain. In addition, the data is not always consistent or sizeable if patients don't follow up, impugning the relevance of the results. Therefore, initiatives in recent years have used social media-based datasets for pregnancy associated health surveillance. In one such instance, Choudhary *et al.* leveraged twitter data to identify postpartum behavioral stance of women from tweet linguistic expression achieving 71% accuracy [18]. In another study, Chandrashekar *et al.* clubbed ADR analysis with pregnancy health surveillance [19]. They used tweets to develop a learning model for identifying cohorts of pregnant women and analyze their respective drug-associated issues drafted over a longitudinal timeline. Working on a similar task Huang *et al.* furnished a pregnancy based tweet corpus using partial tree kernel model to distinguish between true pregnancy cases versus casual mentions [4]. In this study, we primarily gather statistics on legitimate pregnancy cases mentioned on Twitter and further use such instances to analyze and predetermine physical and behavioral health issues expressed by users.

3   Pregnancy Identification Using an Ensemble Neural Network

Previous models in pregnancy case identification have focused primarily on semantic features, or rule-based features, which often fare better [6, 18, 19]. However, stringent lexicon-based learning models are less effective in distinction between false negative and true negative cases. In this paper, we proposed an ensemble neural network model to recognize pregnancy case from tweets via multi-feature fusion. Figure 1 displays the system architecture of our proposed method, which comprises of four key components: tweet embedding, Recurrent Neural Network (RNN)-based feature representation, Convolution Neural Network (CNN)-based feature representation, ensemble feature representation, and SVM classification. We use embedding representation gathered in Tweet Embedding to generate multi layered fused representation from ensemble implementation of RNN and CNN models in tiered fashion. Afterward, the ensemble features are used with SVM for developing the detection model. Each component in the detection model is elucidated in the below subsections.

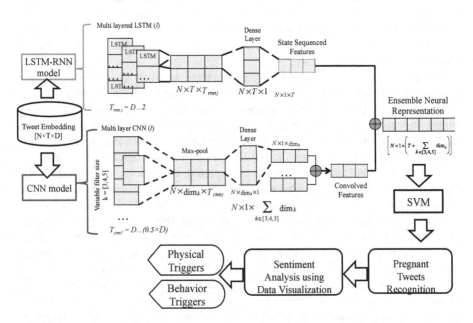

Fig. 1. System architecture for pregnancy identification task

3.1   Tweet Embedding

Tweets are informal communication stuffed within a limited character size making them noisy and slang prone. Therefore, we omit stop words, URLs from our tweet corpus. To represent our tweets in deep learning representation, we adapted Glove embedding corresponding to the tokens in each tweet. However, tweets often have collective or abbreviated words such as hash-tags e.g. "#pregnancyproblems or #safehealthysleep" etc. Given the fair frequency of such lexical representations in our corpus and absence of corresponding vector equivalents for such collectives in Glove, we developed an extended embedding for these words and abbreviations. Using a recursive routine, we identified longest complete word groups per iteration and merged their respective embedding to generate an extended embedding for collective words. In this way, entire candidate tweet dataset was converted to a corresponding pre-trained embedding tensor W of dimension $[N \times T \times D]$, where N $\sim$ total number of candidate tweets, T $\sim$ normalized maximum size of each tweet and D $\sim$ embedding feature dimension (set as 300).

3.2   Representation Learning Using an Ensemble Neural Network Model

We combined a tandem implementation of RNN and CNN models to determine representation features from both. The proposed method is able to extract convolved and sequential lexical features over the embedding tensor. We explain the process of the proposed ensemble learning representation in detail as following.

Recurrent Neural Network-based Representation Features

RNN fares well in learning over long sequenced states of information without running into the problem of vanishing gradients, which is eminent in vanilla RNN. However, such a model falls short while learning contextual dependencies in natural text spread over long durations. To ensure that the model can determine vast longitudinal association between terms in a tweet, we used Long Short-Term Memory network (LSTM) to learn the text representation [20]. LSTM cell has the same inner workings of a RNN model with an addition of logic gate called "forget gate" to withstand the issue of vanishing (or exploding) gradients which enables propagation of learning and prevents error quotient saturation when output for a layer $\sim$ 0 (or 1).

An LSTM cell determines a hidden state representation z_t by mapping each input vector x_t to respective output vector o_x via a series of gates and memory cell defined in below equations:

$$i_t = \sigma(W_{iz}z_{t-1} + W_{ix}x_t + b_i) \tag{1}$$

$$f_t = \sigma\left(W_{fz}z_{t-1} + W_{fx}x_t + b_f\right) \tag{2}$$

$$\tilde{C}_t = tanh(W_{cz}z_{t-1} + W_{cx}x_t + b_c) \tag{3}$$

$$C_t = i_t \odot \tilde{C}_t + f_t \odot C_{t-1} \tag{4}$$

$$o_t = \sigma(W_{oz}z_{t-1} + W_{ox}x_t + b_o) \tag{5}$$

$$z_t = o_t \odot \tanh(C_t) \tag{6}$$

where W denotes weight matrices and b represents bias term. Variables i_t, f_t and o_t represent input gate, forget gate and output gate respectively. $\tilde{C}_t$ and C_t are cell input and cell output activation vectors. σ and $\odot$ stand for sigmoid and element-wise multiplication functions.

Embedding $\sum_{i=1}^{N} W_{i \times T \times D}$ for each tweet is sent through multi layered LSTM cell to generate sequenced representation of dimension $\sum_{i=1}^{N} lstmR_{i \times T \times T_{rnn_l}}$, where $T_{rnn} \sim$ a set of l_f layers of LSTM cells with each layer size T_{rnn_l} described by:

$$T_{rnn_l} = \begin{cases} D, l = 1 \\ (T_{rnn_{l-1}} - (0.25 \times T_{rnn_{l-1}})), \; (1 < l \leq l_f) \wedge (T_{rnn_{l-1}} > 2) \end{cases} \tag{7}$$

The final layer l_f of representation tensor $lstmR$ is reduced by a sigmoid activation dense layer rendering final shape of dimension $\sum_{i=1}^{N} lstmR_{i \times 1 \times T}$. This maps each tweet to a corresponding vector of size $1 \times T$ providing a substitutive linear representation of a tweet from a sequenced contextual association.

Convolution Neural Network-based Representation Features

Image data recognition is a legacy application of Convolution Neural Network (CNN), where a convolving window moving through a matrix of features helps determine submatrix of an image, which can be identified into a particular group even if entire image doesn't present an accurate match [21]. Kim using pre-trained word embedding has explored the same principle involving CNN for sentence classification task [22]. We have employed the same model for our convolving window-based study using a variable window size and multiple convolution filter layers within a window size.

For this CNN-based feature identification, we use the pre-trained tweet embedding (W) described in Sect. 3.1. The embedding tensor $\sum_{i=1}^{N} W_{i \times T \times D}$ is composed of single tweet embedding t_i with token dimension D and normalized/padded tweet size T. Padded tweet embedding in k-word window size for word index j used in feature identification is represented as:

$$w_j = [X_{i_j} \oplus X_{i_{j+1}} \oplus \ldots, \oplus X_{i_{j+k-1}}] \forall X_i \approx \sum_{i=1}^{N} W_{i \times T \times D} \tag{8}$$

where $\oplus$ is the concatenation operator. A convolution window $m_{T_{cnn_l},k} \in R^{T_{cnn_l} \times k \times D}$ with window size $k \in K$ (set of window sizes) and T_{cnn}, a set of l_f layers of CNN where each layer size T_{cnn_l} described by:

$$T_{cnn_l} = \begin{cases} D, l = 1 \\ (T_{cnn_{l-1}} - (0.25 \times T_{cnn_{l-1}})), & (1 < l \le l_f) \wedge (T_{cnn_{l-1}} \le (0.5 \times D)) \end{cases} \tag{9}$$

The convolution window is run over each tweet embedding t_i to generate a feature vector given over per window size k as:

$$c_{j,T_{cnn_l},k} = f\left(w_j \odot m_{T_{cnn_l},k} + b\right) \tag{10}$$

where $\odot$ is an element-wise product, b is the bias term f is a ReLU function. This produces a feature map defined as:

$$c_{T_{cnn_l},k} = [c_{1,T_{cnn_l},k} \oplus c_{2,T_{cnn_l},k} \oplus \ldots, \oplus c_{T-k+1,T_{cnn_l},k}] \tag{11}$$

After convolution completion of each layer l for window size k, max pooling is applied to the output to identify maximum value; $\hat{c}_{Tr_l,k} = \max(c_{Tr_l,k})$. This layered iteration on convolution is continued up to the last layer $l_f \le 0.5 \times D$ where final representation for window size k is resolved giving $\hat{c}_k$ with dimensions $[N \times (T + l_f(1-k)) \times T_{cnn_l}]$. Afterwards a fully connected layer with sigmoid activation is applied to reduce the convolved features into tensor of dimension $[N \times 1 \times dim_k]$ where $dim_k = (T + l_f(1-k)), k \in [K]$. This gives a linear representation of the tweet data for convolution window size k. Upon iterative convolution using variable window size $k \in K \approx [3, 4, 5]$, the corresponding feature map representations for all the given window sizes are concatenated together to form a merged feature block given by:

$$cnnR = \{\hat{c}_{k-1} \oplus \hat{c}_k, 3 \le k - 1 \le 4 \tag{12}$$

where final representation tensor using convolved features has dimensions $\left[N \times 1 \times \sum_{k \in [3,4,5]} dim_k\right]$.

Ensemble Feature Representation

Feature vectors corresponding to each tweet from LSTM-RNN and CNN models are concatenated together to form a composite dense feature representation given by:

$$ensembleR = lstmR \oplus cnnR \tag{13}$$

where representation tensor dimension are $[N \times 1 \times (T + \sum_{k \in [3,4,5]} dim_k)]$. This representation generates a feature vector representation for each tweet. Both train and test representations are generated using ensemble representation form.

3.3 Support Vector Machine-Based Classification

Support vector machines (SVM) have been effective in text categorization especially with lexical feature-based representation often outperforming Naïve Bayes [23]. We used SVM for feature classification with our ensemble deep learning features obtained in Sect. 3.2. SVM splits the high dimensional feature space into separate class sections using a hyper-plane. We used a non-linear mapping kernel i.e. radial bias function (RBF) to project data points into linear feature space.

4   Experimental Results

4.1 Dataset

Pregnancy case detection is a relatively new task in social media related studies with recent works by Chandrashekar *et al.* to study ADRs during pregnancy and by Huang et al. to identify accounts of pregnant women on social media [4, 19]. The prior study developed a relatively small data corpus with 1200 tweets. Therefore, we employed tweet corpus developed by Huang *et al.* consisting of a relatively improved data size with 3000 tweets. The tweets were annotated as 'positive' or 'negative' depending upon the relevance of a tweet to legitimate pregnancy case. There are 642 positive and 2358 negative instances in this corpus, respectively.

4.2 Experimental Setting

To derive credible evaluation results, we employed 10-folds cross validation method. For class imbalance, we employed additional data stratification via repeated sampling of negative cases to match the size of positive within training data when positive cases are less than negative cases. The iterative sampling is repeated until each case type instance is exhaustively used in the model. The evaluation metrics used to determine relative effectiveness of our method include precision, recall, and F_1-score. In addition, we used Glove[5] embedding for our feature representation modeled on Keras[6] platform. We employed SVM function from Scikit-learn SVM library[7]. For representing sentiment-specific lexical feature we use an online word cloud API[8].

4.3 Performance Evaluation of Pregnancy Identification

In Table 1, we compared the effectiveness of our model against different baselines using both linear classifiers as well as deep learning approaches like RNN and CNN. Both Naïve Bayes (NB) and LibShortText[9] (LST) employed *n*-gram model-based

---

[5] https://nlp.stanford.edu/projects/glove/.

[6] https://github.com/keras-team/keras.

[7] https://github.com/scikit-learn/scikit-learn/tree/master/sklearn/svm.

[8] http://nlp.tmu.edu.tw/wordcloud/wordcloud.html.

[9] https://www.csie.ntu.edu.tw/~cjlin/libshorttext/.

features, using statistical frequencies to identify respective feature vectors (bag-of-words, BOW). LST achieved 82% F_1-Score while NB identified only 68% of the target. Table 1 also shows the performance of the state-of-the-art method developed by Huang *et al.* [4], which was based on tree kernel (TK)-SVM (convolution and smoothed, CTK and SPTK). Evident from Table 1, both the tree kernel-based methods had better F_1-scores compared to NB in determining positive cases, however the lexical structure overlap across positive and negatives instances compounded with class size imbalance resulted in low recall on the positive cases.

Finally, we access the performance of the other standalone deep neural network (DNN) models like LSTM-RNN and CNN in comparison to our approach. Evident from Table 1, sequenced features of LSTM-RNN fare better in identifying pregnancy cases as opposed to CNN baseline implementation. Our ensemble neural representation works further effectively in screening out false positives, developing 95% efficiency in predicting true instances.

Our model used an aggregate of sequenced and convolved features for elaborate tweet representation. Compared to models using statistical or syntactic features, our ensemble features register moderately high performance in distinguishing against the negative instances. Test case study initiated in this work is 95% effective in determining legitimate pregnancy cases on Twitter. The proposed method can be leveraged to collect statistics on active pregnancies cases and provide additional health care support to women as briefed in further analysis.

Table 1. Summary of ensemble-representation SVM model against baselines

Features	Method	Positive	Negative	avg_μ
		Precision /Recall /F1-score		
BOW	NB	0.31/0.48/0.38	0.83/0.71/0.77	0.72/0.66/0.68
	LST	0.62/0.47/0.53	0.86/0.92/0.89	0.81/0.82/0.82
TK	CTK	0.68/0.44/0.54	0.86/0.94/0.90	0.82/0.83/0.82
	SPTK	0.71/0.46/0.56	0.86/0.95/0.90	0.83/0.84/0.83
DNN	CNN	0.23/0.86/0.36	0.85/0.21/0.34	0.71/0.35/0.35
	LSTM-RNN	0.32/0.83/0.47	0.92/0.53/0.67	0.79/0.59/0.63
	Our method	**1.00/0.76/0.95**	**0.94/1.00/0.95**	**0.97/0.86/0.95**

4.4   Sentiment Analysis on Pregnancy Cases

The larger scope of this case study is to exploit social media data in developing medical and behavioral health stats for pregnant women. During the course of their pregnancy, women often tweet about various health concerns related to pregnancy or express exhaustion or depression during the term. Key terms from such tweets can help us identify common health concerns and triggers for behavioral changes women experience during pregnancy. Therefore, as one of the first initiatives known to the best of our knowledge, we extended a sentiment analysis determination on the identified pregnancy related tweets.

Owing to limited size of positively recognized pregnancy cases, we decided to proceed with an exploratory study identifying and classifying sentiments in pregnancy tweets. Considering the scope of our work and in alignment with the most recurring themes in tweets, we settled on four sentiment classes based on a manual sentiment annotation of the identified tweets shown in Table 2.

Table 2. Pregnancy associated sentiment definition

Sentiment	Definition	#Tweet
Neutral (N)	Tweets which don't provide any health related or behavioral information on pregnancy	350
Happy (H)	Tweets with inference of happiness or excitement related to pregnancy e.g. happy, can't wait to see my baby etc	73
Distress (D)	Tweets which express any health related concern e.g. nausea, swollen feet etc	62
Annoyed (A)	Tweets which infer negative behavioral emotions e.g. sad, depressed, exhausted etc	157

Our analysis identified about 62 tweets discussing health issues and 157 tweets discussing negative behavioral emotions shown in Table 2. To identify trigger terms for respective sentiments we studied sentiment-lexical feature co-occurrence using Log-Likelihood Ratio (LLR) method. Lexical terms with higher log likelihood to occur with a particular sentiment class were ranked higher in the corresponding group. Using LLR score for each word, a word cloud was formed to showcase key terms identified in the selective group.

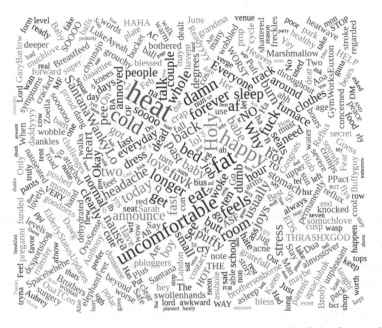

Fig. 2. Word cloud generated from pregnancy sentiment keywords (Color figure online)

Red, blue and green colored terms represent sentiments 'Distress', 'Annoyed' and 'Happy' respectively in the word cloud Fig. 2. Cursory analysis of the terms from LLR based study reveals depression instances expressed by use of terms like feeling '*fat*', '*uncomfortable*' in their appearance or phrases like 'forever *pregnant*' or '*damm*' to identify their exhaustion. We also identified some of the most common physical symptoms women experience during their term like 'catching *cold*', 'sensitive to *heat*', 'prolonged *backaches*', 'frequent *headaches*' and 'feeling *puffy*'.

5 Conclusion and Future Work

In this study, we have explored an ensemble representation model to identify most elaborate features from smaller learning instances like 'tweets'. Our ensemble neural network-based representation can identify true pregnancy accounts discussed on social media effectively. With an overall goal to leverage social media for learning health stats on pregnant women, we also performed an exploratory sentiment analysis study on pregnant women accounts, which is new to pregnancy-related social media studies. Using LLR based feature selection method, we identified and ranked terms associated with various recurring sentiments discussed in tweets like diseases, symptoms and different behavioral issues, which can help identify triggers to physical and behavioral conditions beneficial for treatment in clinical practices.

The small size of the dataset has been a consistent limitation for sentiment analysis study in our task. Therefore, in the future, we intend to expand our exploratory study with more diverse datasets over extended timelines for exhaustive monitoring of medical changes women undergo during pregnancy. We also intend to do a multi-annotator based tweet-sentiment study to automate learning of sentiments. The initial results of our study are quite promising and stable and warrant further research.

Acknowledgments. We are grateful to the anonymous reviewers for their insightful comments. This research was supported by the Ministry of Science and Technology of Taiwan under grant number MOST 106-2218-E-038-004-MY2, MOST 107-2634-F-001-005, and MOST 107-2319-B-400-001.

References

1. Woodall, J., Calisher, C.H.: ProMED-mail: background and purpose. Emerg. Infect. Dis. **7**(3 Suppl.), 563 (2001)
2. Mykhalovskiy, E., Weir, L.: The global public health intelligence network and early warning outbreak detection: a Canadian contribution to global public health. Can. J. Public Health **97** (1), 42–44 (2006)
3. Brownstein, J.S., Freifeld, C.C., Madoff, L.C.: Digital disease detection — harnessing the web for public health surveillance. New England J. Med. **360**(21), 2153–2157 (2009). https://doi.org/10.1056/NEJMp0900702
4. Huang, Y., et al.: Incorporating dependency trees improve identification of pregnant women on social media platforms. In: 2017 Proceedings of the International Workshops on Digital Disease Detection using Social Media (DDDSM-2017), Taipei, pp. 26–32 (2017)

5. Gomide, J.: Dengue surveillance based on a computational model of spatio-temporal locality of Twitter. In: Proceedings of the 3rd International Web Science Conference, p. 3. ACM (2011)

6. Diaz-Aviles, Stewart, A.: Tracking Twitter for epidemic intelligence: case study: Ehec/hus outbreak in Germany. In: 2011 Proceedings of the 4th Annual ACM Web Science Conference, pp. 82–85. ACM (2012)

7. Odlum, M.: How Twitter can support early warning systems in Ebola outbreak surveillance. In: 143rd APHA Annual Meeting and Exposition, 31 October–4 November 2015. APHA (2015)

8. McGough, S.F., Brownstein, J.S., Hawkins, J.B., Santillana, M.: Forecasting zika incidence in the 2016 latin america outbreak combining traditional disease surveillance with search, social media, and news report data. PLoS Negl. Trop. Dis. **11**(1), e0005295 (2017). https://doi.org/10.1371/journal.pntd.0005295

9. Mejova, Y., Haddadi, H., Noulas, A., Weber, I.: # foodporn: obesity patterns in culinary interactions. In: Proceedings of the 5th International Conference on Digital Health 2015, pp. 51–58. ACM (2015)

10. Aphinyanaphongs, Y., Ray, B., Statnikov, A., Krebs, P.: Text classification for automatic detection of alcohol use-related tweets. In: International Workshop on Issues and Challenges in Social Computing (2014)

11. Leaman, R., Wojtulewicz, L., Sullivan, R., Skariah, A., Yang, J., Gonzalez, G.: Towards internet-age pharmacovigilance: extracting adverse drug reactions from user posts to health-related social networks. In: Proceedings of the 2010 Workshop on Biomedical Natural Language Processing, pp. 117–125 (2010)

12. Bian, J., Topaloglu, U., Yu, F.: Towards large-scale twitter mining for drug-related adverse events. In: Proceedings of the 2012 International Workshop on Smart Health and Wellbeing, pp. 25–32. ACM (2012)

13. Sarker, A., Gonzalez, G.: Portable automatic text classification for adverse drug reaction detection via multi-corpus training. J. Biomed. Inf. **53**, 196–207 (2015)

14. Dai, H.J., Touray, M., Jonnagaddala, J., Syed-Abdul, S.: Feature engineering for recognizing adverse drug reactions from Twitter posts. Information **7**(2), 27 (2016). https://doi.org/10.3390/info7020027

15. Banjari, I., Kenjeri, D., Šolić K., Mandić, M.L.: Cluster analysis as a prediction tool for pregnancy outcomes. Collegium Antropol. **39**(1), 247–252 (2015)

16. Laopaiboon, M.: Imezoglu. Advanced maternal age and pregnancy outcomes: a multicountry assessment. BJOG: Int. J. Obstet. Gynaecol. **121**(s1), 49–56 (2014)

17. Wettach, C., Thomann, J., Lambrigger-Steiner, C., Buclin, T., Desmeules, J., von Mandach, U.: Pharmacovigilance in pregnancy: adverse drug reactions associated with fetal disorders. J. Perinat. Med. **41**(3), 301–307 (2013)

18. De Choudhury, M., Counts, S., Horvitz, E.: Predicting postpartum changes in emotion and behavior via social media. In: Proceedings of the SIGCHI Conference on Human Factors in Computing Systems, pp. 3267–3276. ACM (2013)

19. Chandrashekar, P.B., Magge, A., Sarker, A., Gonzalez, G.: Social media mining for identification and exploration of health-related information from pregnant women (2017). CoRR, abs/1702.02261

20. Hochreiter, S., Schmidhuber, J.: Long short-term memory. Neural Comput. **9**(8), 1735–1780 (1997). https://doi.org/10.1162/neco.1997.9.8.1735

21. Lecun, Y., et al.: Comparison of learning algorithms for handwritten digit recognition. In: Fogelman, F., Gallinari, P. (eds.) International Conference on Artificial Neural Networks, Paris, pp. 53–60. EC2 & Cie (1995)

22. Kim, Y.: Convolutional neural networks for sentence classification. In: Proceedings of the 2014 Conference on Empirical Methods in Natural Language Processing, EMNLP 2014, October 2014, Doha, Qatar, A meeting of SIGDAT, a Special Interest Group of the ACL pp. 1746–1751 (2014)
23. Joachims, T.: Text categorization with Support Vector Machines: Learning with many relevant features. In: Nédellec, C., Rouveirol, C. (eds.) ECML 1998. LNCS, vol. 1398, pp. 137–142. Springer, Heidelberg (1998). https://doi.org/10.1007/BFb0026683

Chinese Governmental Named Entity Recognition

Qi Liu[1,2], Dong Wang[1,2(✉)], Meilin Zhou[1,2], Peng Li[1,2], Baoyuan Qi[1,2],
and Bin Wang[1,2]

[1] Institute of Information Engineering Chinese Academy of Sciences, Beijing, China
{liuqi,wangdong,zhoumeilin,lipeng,qibaoyuan,wangbin}@iie.ac.cn
[2] University of Chinese Academy of Sciences, Beijing, China

Abstract. Named entity recognition (NER) is a fundamental task in natural language processing and there is a lot of interest on vertical NER such as medical NER, short text NER etc. In this paper, we study the problem of Chinese governmental NER (CGNER). CGNER serves as the basis for automatic governmental text analysis, which can greatly benefit the public. Considering the characteristics of the governmental text, we first formulate the task of CGNER, adding one new entity type, i.e., policy (POL) in addition to the generic types such as person (PER), location (LOC), organization (ORG) and title (TIT) for recognition. Then we constructed a dataset called GOV for CGNER. We empirically evaluate the performances of mainstream NER tools and state-of-the-art BiLSTM-CRF method on the GOV dataset. It was found that there is a performance decline compared to applying these methods on generic NER dataset. Further studies show that compound entities account for a non-negligible proportion and using the classical BIO (Begin-Inside-Outside) annotation cannot encode the entity type combination effectively. To alleviate the problem, we propose to utilize the compound tagging and BiLSTM-CRF for doing CGNER. Experiments show that our proposed methods can significantly improve the CGNER performance, especially for the LOC, ORG and TIT entity types.

Keywords: Named entity recognition · E-government
Information extraction

1 Introduction

Named entity recognition (NER) refers to the task of detecting concepts from textual data and classifying the concepts into predefined categories such as persons, organizations etc. The categories, named as entity types, are domain dependent. Although the study for NER in the generic domain such as journalistic articles is abundant [1], the study is scarce when it comes to vertical NER, since text characteristics and extraction demands are quite different for different domains [2].

© Springer Nature Switzerland AG 2018
Y.-H. Tseng et al. (Eds.): AIRS 2018, LNCS 11292, pp. 16–28, 2018.
https://doi.org/10.1007/978-3-030-03520-4_2

Nowadays, more and more countries are putting emphasis on the e-government construction. Texts related to government policies, government decisions and government actions are quickly generated. The demands for governmental text analysis exist widely for different users such as governments, news agencies and ordinary people. We define the NER task for the government related texts as governmental NER. Especially, we focus on Chinese governmental NER, i.e. CGNER for short, in this paper. Although CGNER is in great demand, we surprisingly found that there is very few studies reported in the literature thus far.

As a new problem, the specialties of CGNER are reflected in the following two aspects:

- There is an additional extraction demand. Government policy (POL) is one kind of new entity type. Compared to English governmental texts, the POL entities in Chinese ones tend to be created by grouping Chinese characters, such as "三支一扶" (Three Supports and One Assistance), "营改增" (Business Tax Replaced with Value-added Tax) etc. These policy entities can either be from the central government or from the grassroots governments. Also with the development of government activities, new policy entities are created accordingly.
- The entity composition is special. For Chinese governmental texts, the title (TIT), location (LOC) and organization (ORG) entities are usually compound entities. For example, an ORG entity "海淀区民政局" comprises of one LOC entity "海淀区" and one ORG entity "民政局". Since there are many possible combinations, the compound entities are very sparse in the text, i.e., there are very little texts containing the same compound entity. The diversity for the same prefix entity under different contexts also increases the difficulty of recognition. For example, "北京海淀区区长" (TIT) and "北京海淀区民政局" (ORG) both share the same entity prefix "北京海淀区", however, their final entity types are different which makes the tag prediction of the prefix entity characters difficult. A comparison of the texts from the governmental domain and the generic domain is shown in Fig. 1.

(a)   (b)

Fig. 1. (a) An example text from governmental domain(GOV dataset). (b) An example text from generic domain(MARS dataset).

Since there is no public dataset available for CGNER, we constructed a dataset called GOV for method testing. Specifically, we crawled news from two kinds of official websites: the official central media websites such as Xinhua Net[1] and the official government websites such as the Central People's government Net[2]. To validate the difficulty of CGNER, we conducted three experiments: (1) We directly applied the mainstream NER tools on the GOV data without training and found that the performance is quite poor. (2) We trained a model on the dataset MSRA which corresponds to the generic domain and tested it on GOV data. We found that there is a large performance decline for the LOC and ORG entity types. (3) we apply the state-of-the-art BiLSTM-CRF method on GOV dataset and show the performance improvement. These experiments just indicate as a new domain, CGNER should be treated specially.

To further improve the performance, we did analysis on the entity composition and found that compound entities account for a non-negligible proportion, nearly 50%. The classical BIO (Begin-Inside-Outside) annotation cannot capture the information of entity type combination effectively. To alleviate the problem, we propose a new tagging scheme called compound tagging for character labeling. For example, using compound tagging, the tag for "北" in "北京海淀区区长" is B_LOC_B_TIT while the tag for "北" in "北京海淀区民政局" is B_LOC_B_ORG.

Our contributions are as follows:

- To the best of our knowledge, we are the first to define the task of CGNER and build a corpus for method validation (Sect. 3);
- We empirically evaluate the performances of mainstream NER tools and state-of-the-art BiLSTM-CRF method on the GOV dataset (Sect. 4);
- We propose to use compound tagging and BiLSTM-CRF for doing CGNER, which significantly outperforms the above baselines (Sect. 5).

2   Related Work

As a fundamental task, NER has been well studies. The most related works to our paper are as follows.

2.1   Methods Based on Statistical Machine Learning

These approaches treat the entity recognition as the problem of classification or sequential labeling, i.e., predicting the word label by leveraging the context features. Specifically, classification methods such as Logistic Regression, Decision Trees [3], Support vector machine [4], Maximum entropy [5] have all been proposed in earlier conferences. The most popular method for sequential labeling in NER is HMM [6] and CRF [7], which predicts the word labels not only using the context word features but also using the context word labels. In fact, CRF has been taken as the standard method in mainstream NER tools. As for the

---

[1] http://www.xinhuanet.com/.
[2] http://www.gov.cn/.

traditional statistical NER, the most difficult part is to manually define the features, which is also costly and experience dependent. Recently, some rule-based approaches [8] are proposed to solve NER, but it's time-consuming to define enough good rules to cover all situations.

2.2 Methods Based on Neural Networks

With the development of deep learning, neural network has become practical in NER. Specifically, Collobert [9] proposed an unified framework for sequence labeling and Huang [10] proposed to combine LSTM and CRF for doing NER, i.e., BiLSTM-CRF. The BiLSTM-CRF quickly became the standard method for neural network based NER and several methods were further proposed for improvement such as [11–15]. The advantage of using neural network is that people do not have to define features, which are learned from data automatically.

2.3 Methods for Compound Named Entity Recognition

Compound named entity refers to those entities with sub entities. The standard NER methods are designed to extract entities through one pass mode which cannot identify compound entities effectively. One typical method for this problem is called two-stage CRF [16], whose idea is to recognize unit entities in the first stage, and then convert the results into features to identify compound entities in the second stage. Though it is a popular method, its performance can be easily affected by the error propagation under this pipeline mode.

3 Defining the Task of CGNER

3.1 Task

The task of CGNER focuses on extracting entities of five types: person (PER), location (LOC), organization (ORG), title (TIT) and policy (POL). The first three types are the common entity types. The TIT type is also defined in the GATE system[3]. The last type POL is unique for CGNER. The interpretation and the examples of five entity types are provided in Table 1.

The evaluation of CGNER is the same as traditional NER. The evaluation measures are precision (P), recall (R) and F-measure (F1).

Table 1. The interpretation and examples of five entity types

Entity Type	Interpretation	Examples
PER	person name etc.	李克强
LOC	cities, districts, streets etc.	海淀区中关村
ORG	companies, agencies, institutions, etc.	海淀区民政局
TIT	official position etc.	海淀区区长
POL	policy name etc.	一带一路 , 双创

---

[3] https://gate.ac.uk/projects.html.

3.2   Dataset

Since there is no public dataset available for CGNER, we have to build the dataset from scratch. To make the texts representative, we used a web crawler to download web pages from three authority governmental websites, which are the websites of Central People's government, the People's Daily Online[4] and the Xinhua net. The crawl lasted from June 2017 to September 2017.

After that, we have collected about 60,000 web pages. We then extracted their main contents and randomly selected 8,000 sentences for experiments. These sentences were annotated separately by two students, and then integrated by the third student. These sentences contain 427,860 characters and 8,191 entities in total, which corresponds to 53.5 characters and 1.02 entities for each sentence on average. The number of sentences for NER is similar to other vertical NER corpuses [2]. We call the dataset GOV in the paper. Table 2 presents the statistical result of GOV.

We manually labeled the sentences using the classical BIO annotation scheme, where B and I represent the begin and inside of an entity respectively while O represents non-entity part. More specifically, each character in a sentence is labeled using a tag which corresponds to a combination of position and entity types. The position is one of B/I/O and the entity type is one of the five entity types. Under this annotation, there are $2 * 5 + 1 = 11$ kinds of tags. An example of BIO annotation is shown in Table 8.

Table 2. The quantity, ratio and average length of entities in the dataset

	Quantity	Ratio	Average length
PER	1,697	20.72%	2.81
LOC	1,957	23.89%	2.91
ORG	1,729	21.11%	4.71
TIT	1,010	12.33%	7.20
POL	1,798	21.95%	4.97
Total	8,191	100%	4.52

4   Empirical Studies on CGNER

In this part, we empirically test the mainstream tools and the state-of-the-art method for NER on the Chinese governmental texts.

---

[4] http://www.people.com.cn/.

4.1   Testing NER Tools on GOV

Since NER is a well studied problem, there are several tools with trained models which can be used directly. The most representative tools for Chinese NER are StanfordNER[5] and LTP[6]. Both of them use the CRF model though they are trained using different data. We test these two tools on GOV to see whether they can meet the requirement. The evaluation results are presented in Table 3.

Table 3. F_1 measures of applying StanfordNER and LTP on GOV

Tools	PER	LOC	ORG
StanfordNER	86.56	72.44	67.47
LTP	88.21	73.16	68.71

From the table, we can see that the TIT and POL are not supported by the current tools. The performances without training on governmental texts are not satisfactory, especially for LOC and ORG entities. There is still much room for improvement.

4.2   Training BiLSTM-CRF Model on Generic Domain and Testing on GOV

In this part, we test the state-of-the-art method called BiLSTM-CRF for CGNER. Similar to Sect. 4.1, we still train the model on a generic dataset and test it on GOV. If the testing performances of the trained model on the generic domain and GOV are of small difference, then it means domain difference is unimportant. Otherwise, the characteristics of governmental texts are different and CGNER should be studied specially.

Data Preparation. We used the MSRA dataset in 3rd SIGHAN Bakeoff [17] to represent the generic domain. For testing, we randomly selected 3,000 sentences from GOV and MSRA respectively, which are named as GOV-Test and MSRA-Test. For training, we use the left data of MSRA, named as MSRA-Train.

BiLSTM-CRF Model. The BiLSTM-CRF Model is one of the most representative neural network based method for NER. It's a three layer architecture which is shown in Fig. 2.

---

[5] https://nlp.stanford.edu/software/CRF-NER.shtml.
[6] https://www.ltp-cloud.com/.

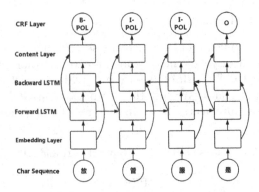

Fig. 2. The architecture of BiLSTM-CRF model for NER

Experiments. We use the following hyper-parameters for BiLSTM-CRF model. The number of LSTM units is 300, no peepholes. The dropout rate is set to 0.5 for regularization. Word embedding is pretrained by gensim[7] using Chinese Wikipedia backup dump, whose dimensions are 200. Adam [18] is adopted as the optimization method. The gradient clipping is set to 5 to prevent gradient explosion. Batch size is set to 64 and the learning rate is set to 0.001.

We present the experimental results in Table 4. From the table, we can see that there is a significant performance drop for LOC and ORG entities from MSRA-Test to GOV-Test. This indicates that the domain difference cannot be ignored and even the BiLSTM-CRF has limited capability for domain transferring.

Table 4. F_1 measures of applying the model trained on MSRA-Train to MSRA-Test and GOV-Test

	PER	LOC	ORG
MSRA-Test	89.57	88.30	82.10
GOV-Test	**85.47**	**75.29**	**69.73**

4.3 Training and Testing BiLSTM-CRF on GOV

In this part, we investigate the performance of BiLSTM-CRF on GOV, where 70% of the data is used for training and 30% is used for testing. For the training data, we also left 10% of the data as the dev data. By applying BiLSTM-CRF, we want to see the CGNER performances of the entity type TIT and POL and compare the recognition difficulties for different entity types. Table 5 presents the evaluation results on dev set and testing set. We can see that our trained neural network does not overfit.

---

[7] https://radimrehurek.com/gensim/.

Table 5. F_1 measures of BiLSTM-CRF on GOV dev and testing set

GOV	PER	LOC	ORG	TIT	POL
dev set	86.97	77.39	72.10	81.62	61.91
testing set	**85.69**	**76.23**	**70.83**	**80.25**	**60.87**

From Table 5, we can get the following conclusions:

- The order of recognition difficulty for the five entity types is: POL > ORG > LOC> TIT> PER. The identification of POL is the most difficult in all five types. As we see from the GOV data, most of the POL entities are condensed words which comprise of several characters. And the context of POL is also diverse compared to the other entity types. The above reasons make it difficult to detect the POL entities.
- The recognition of LOC and ORG entities are also difficult. By viewing the data, we conclude that the difficulty is caused by the phenomena of compound entities, i.e., many entities of LOC and ORG are actually composed of several fine-grained entities.

For further investigation, we compare the proportion of compound entities over MSRA-Test data and GOV-Test data, which are given in Table 6. From the table, we can see that the compound entity proportions for LOC and ORG in GOV are significantly higher than those in MSRA, exceeding more than 30% at most. This can also explain the performance drop of transferring the model learned from MSRA-Train to GOV-Test in Sect. 4.2.

Table 6. The proportion of compound entities of different types in GOV-Test and MSRA-Test

Compound entity type	GOV-Test	MSRA-Test
LOC	35.3%	10.7%
ORG	55.8%	20.2%
TIT	60.2%	-

Table 7. A statistics on the proportions of compound named entities in GOV

	Number of sub entities	Proportion
Entity without sub entity	0	40.3%
Compound entity	1	50.5%
	2	7.4%
	3	1.7%
	Above 3	0.1%

Besides, we calculate the proportions of compound entities with different number of sub entities on GOV data, which are presented in Table 7. From the table, we can see that most compound entities have only one sub entity, i.e. 50.5%; the total proportion for the compound entities is nearly 60%. This indicates that the governmental texts are quite different from the texts of generic domain.

5   Our Proposed Approach for CGNER

Since the compound entities account for a large proportion in governmental texts, we focus on tackling the identification of compound entities in this part.

5.1   The Tagging Scheme

Motivated by the work [19], we propose a new tagging scheme for character in compound entity by packing the character tags in sub entity and compound entity as a new combined tag. We name our tagging scheme as compound tagging (CT in short). Table 8 presents an example of CT. For the compound entity "北京海淀区民政局", the first sub entity "北京" is of LOC type and the complete entity is of ORG type, so the tag for "北" is a combination of two tags, i.e., B_LOC_B_ORG, for which B_LOC corresponds to the tag in sub entity and B_ORG corresponds to the tag in complete entity. For those non-compound entities, the character tags are exactly the same as those in BIO.

Table 8. An illustration of BIO and compound tagging

labeling type	北	京	海	淀	区	民	政	局
BIO	B_ORG	I_ORG	I_ORG	I_ORG	I_ORG	I_ORG	I_ORG	I_ORG
CT	B_LOC_B_ORG	I_LOC_I_ORG	B_LOC_I_ORG	I_LOC_I_ORG	I_LOC_I_ORG	B_ORG_I_ORG	I_ORG_I_ORG	I_ORG_I_ORG

5.2   Our Approach

Our proposed approach for CGNER is to train a BiLSTM-CRF model on the sentences with tags using compound tagging. According to the definition of CT, we manually relabeled the sentences in GOV dataset. We name our method as BiLSTM-CRF+CT.

5.3   Baselines

To validate the effectiveness of our proposed approach, we select Two-Stage-CRF [16] with BIO as our baselines, which we name as 2CRF+BIO. Furthermore, we replace the CRF model with BiLSTM-CRF to formulate a new baseline, which we name as 2BiLSTM-CRF+BIO.

2CRF+BIO. This model utilizes two CRFs for extracting sub entities and compound entities sequentially. More specifically, the tags generated by the first CRF are taken as the input for the second CRF. Since the corresponding tagging scheme is BIO, we name the baseline as 2CRF+BIO.

2BiLSTM-CRF+BIO. On the basis of 2CRF+BIO, we replace CRF with BiLSTM-CRF to formulate this new baseline. The generated tags from the first BiLSTM-CRF will be encoded into one-hot, which will be the input for the second BiLSTM-CRF in conjunction with word embedding. And we name the model as 2BiLSTM-CRF+BIO.

5.4   Experiments

As in Sect. 4.3, 70% of the data is used for training and 30% is used for testing. The hyper-parameters are set the same as in Sect. 4.3. The experimental results are given in Table 9. From the table, we can see 2BiLSTM-CRF+BIO outperforms 2CRF+BIO, and the reason may lie BiLSTM can learn the features automaticly.

Table 9. A comparison of F_1 measures achieved by our approach and the baselines

	PER	LOC	ORG	TIT	POL
2CRF+BIO	84.05	74.94	69.33	78.66	59.92
2BiLSTM-CRF+BIO	86.20	75.36	71.72	80.18	61.52
BiLSTM-CRF+CT	**87.29**	**77.28**	**73.53**	**82.24**	**62.80**

We also find our proposed BiLSTM-CRF+CT is the most effective compared to the other baselines, and there is a significant improvement for the entity type LOC, ORG and TIT. Two factors make the improvement:

- Two baselines both are Two-Stage models, so their performance will be affected by the error propagation under this pipeline mode.
- CT introduces more fine-grained annotation, i.e., more entities types, which increases the complexity of model and improves the expression of model. For example, in CT, TIT comprises of ORG_TIT (e.g., "药管局局长") and LOC_TIT (e.g., "长春市市长"), so model can learn the more specific features for various entities.

In addition, improvement for POL is inferior, which indicates that the POL extraction is still worth studying.

5.5 Case Study

In this section, we pick a specific example to show the advantage of CT. The first row and the second row of Table 10 are true tags annotated in BIO and CT. The last three rows are predicted tags generated by 2CRF+BIO, 2BiLSTM-CRF+BIO and BiLSTM-CRF+CT. Restrained by limited space, we present a CT tag in three rows.

Table 10. An illustration of a specific example

		国	家	食	品	药	品	监	督	管	理	总	局	局	长
true tags	BIO	B.TIT	L.TIT	L.TIT	L.TIT	L.TIT	L.TIT	L.TIT	L.TIT	L.TIT	L.TIT	L.TIT	L.TIT	L.TIT	L.TIT
	CT	B.ORG	L.ORG	L.ORG	L.ORG	L.ORG	L.ORG	L.ORG	L.ORG	L.ORG	L.ORG	L.ORG	L.ORG	B.TIT	L.TIT
		-	-	-	-	-	-	-	-	-	-	-	-	-	-
		B.TIT	L.TIT	L.TIT	L.TIT	L.TIT	L.TIT	L.TIT	L.TIT	L.TIT	L.TIT	L.TIT	L.TIT	L.TIT	L.TIT
predicted tags	2CRF+BIO	B.ORG	L.ORG	L.ORG	L.ORG	L.ORG	L.ORG	L.ORG	L.ORG	L.TIT	L.TIT	L.TIT	L.TIT	L.TIT	L.TIT
	2BiLSTM+BIO	B.ORG	L.ORG	L.ORG	L.ORG	L.ORG	L.ORG	L.ORG	L.ORG	L.ORG	L.ORG	L.ORG	L.ORG	0	0
	BiLSTM-CRF+CT	B.ORG	L.ORG	L.ORG	L.ORG	L.ORG	L.ORG	L.ORG	L.ORG	L.ORG	L.ORG	L.ORG	L.ORG	B.TIT	L.TIT
		-	-	-	-	-	-	-	-	-	-	-	-	-	-
		B.TIT	L.TIT	L.TIT	L.TIT	L.TIT	L.TIT	L.TIT	L.TIT	L.TIT	L.TIT	L.TIT	L.TIT	L.TIT	L.TIT

As our expectation: In CT, "国家食品药品监督管理总局局长", which is combination of ORG and TIT, was annocated correctly; but in BIO, without fine-grained annotation, "国家食品药品监督管理总局" was annocated as ORG incorrectly.

In addition, BiLSTM-CRF+CT can find the compound entities and it's inner entities at the same time, which improves the efficiency.

6 Conclusion

In this paper, we propose the problem of CGNER, i.e., Chinese governmental named entity recognition. First, we analyze the demand of CGNER and incorporate the policy (POL) as a new entity type. Then we construct a dataset called GOV by collecting news from important governmental websites. We empirically validate the performances of the mainstream NER tools and state-of-the-art method for CGNER. It is found that governmental texts are quite different from generic texts and existing NER tools or models trained on generic domain cannot transfer to the governmental domain effectively. Further studies show that compound entities account for nearly 50% in governmental texts. To tackle the problem, we propose to utilize the compound tagging and BiLSTM-CRF for doing CGNER. Experiments demonstrate the effectiveness of our approach.

Acknowledgment. We would like to thank the anonymous reviewers for their insightful comments and suggestions. This research is supported by the The National Key Research and Development Program of China (grant No. 2016YFB0801003 & 2017YFB0803301).

References

1. Grishman, R., Sundheim, B.: Message understanding conference-6: a brief history. In: COLING 1996 Volume 1: The 16th International Conference on Computational Linguistics, vol. 1 (1996)
2. Uzuner, Ö., South, B.R., Shen, S., DuVall, S.L.: 2010 i2b2/va challenge on concepts, assertions, and relations in clinical text. J. Am. Med. Inf. Assoc. **18**(5), 552–556 (2011)
3. Sekine, S.: Description of the Japanese ne system used for met-2. In: Seventh Message Understanding Conference (MUC-7): Proceedings of a Conference Held in Fairfax, Virginia, 29 April–1 May 1998 (1998)
4. Asahara, M., Matsumoto, Y.: Japanese named entity extraction with redundant morphological analysis. In: Proceedings of the 2003 Conference of the North American Chapter of the Association for Computational Linguistics on Human Language Technology, vol. 1, pp. 8–15. Association for Computational Linguistics (2003)
5. Borthwick, A., Sterling, J., Agichtein, E., Grishman, R.: Description of the MENE named entity system as used in MUC-7. In: Proceedings of the Seventh Message Understanding Conference (MUC-7), Fairfax, Virginia, 29 April–1 May 1998 (1998)
6. Bikel, D. M., Miller, S., Schwartz, R., Weischedel, R.: Nymble: a high-performance learning name-finder. In: Proceedings of the fifth conference on Applied natural language processing, pp. 194–201. Association for Computational Linguistics (1997)
7. McCallum, A., Li, W.: Early results for named entity recognition with conditional random fields, feature induction and web-enhanced lexicons. In: Proceedings of the seventh conference on Natural language learning at HLT-NAACL 2003, vol. 4, pp. 188–191. Association for Computational Linguistics (2003)
8. Yao, Y., Sun, A.: Mobile phone name extraction from internet forums: a semi-supervised approach. World Wide Web **19**(5), 783–805 (2016)
9. Collobert, R., Weston, J., Bottou, L., Karlen, M., Kavukcuoglu, K., Kuksa, P.: Natural language processing (almost) from scratch. J. Mach. Learn. Res. **12**(Aug), 2493–2537 (2011)
10. Huang, Z., Xu, W., Yu, K.: Bidirectional LSTM-CRF models for sequence tagging. arXiv preprint arXiv:1508.01991 (2015)
11. Chiu, J.P., Nichols, E.: Named entity recognition with bidirectional LSTM-CNNs. arXiv preprint arXiv:1511.08308 (2015)
12. Ma, X., Hovy, E.: End-to-end sequence labeling via bi-directional LSTM-CNNs-CRF. arXiv preprint arXiv:1603.01354 (2016)
13. Lample, G., Ballesteros, M., Subramanian, S., Kawakami, K., Dyer, C.: Neural architectures for named entity recognition. arXiv preprint arXiv:1603.01360 (2016)
14. Dong, C., Zhang, J., Zong, C., Hattori, M., Di, H.: Character-based LSTM-CRF with radical-level features for chinese named entity recognition. In: Lin, C.-Y., Xue, N., Zhao, D., Huang, X., Feng, Y. (eds.) ICCPOL/NLPCC -2016. LNCS (LNAI), vol. 10102, pp. 239–250. Springer, Cham (2016). https://doi.org/10.1007/978-3-319-50496-4_20
15. Dernoncourt, F., Lee, J.Y., Szolovits, P.: Neuroner: an easy-to-use program for named-entity recognition based on neural networks. arXiv preprint arXiv:1705.05487 (2017)
16. Fu, C., Fu, G.: Morpheme-based chinese nested named entity recognition. In: 2011 Eighth International Conference on Fuzzy Systems and Knowledge Discovery (FSKD), vol. 2, pp. 1221–1225. IEEE (2011)

17. Levow, G.-A.: The third international chinese language processing bakeoff: word segmentation and named entity recognition. In: Proceedings of the Fifth SIGHAN Workshop on Chinese Language Processing, pp. 108–117 (2006)
18. Kingma, D.P., Ba, J.: Adam: a method for stochastic optimization. arXiv preprint arXiv:1412.6980 (2014)
19. Zheng, S., Wang, F., Bao, H., Hao, Y., Zhou, P., Xu, B.: Joint extraction of entities and relations based on a novel tagging scheme. arXiv preprint arXiv:1706.05075 (2017)

StanceComp: Aggregating Stances from Multiple Sources for Rumor Detection

Hao Xu[✉] and Hui Fang

Department of Electrical and Computer Engineering, University of Delaware,
Newark, USA
{haoxu,hfang}@udel.edu

Abstract. With the rapid development of the Internet, social media has become a major information dissemination platform where any users can post and share information. Although this facilitates the share of breaking news, it also becomes the fertile land for the spread of malicious rumors. On the contrary, online news media might lag behind on reporting breaking news but their articles are more reliable since the journalists often go to verify the information before they report it. Intuitively, when users try to decide whether to trust a claim they saw on the social media, they would want to check stances of the same claim on social media and news media. More specifically, they want to know the opinions of other users, i.e., whether they support or against the claim. To facilitate such a process, we develop *StanceComp*(https://stancecomp.herokuapp.com), which aggregates the relevant information about a claim and compares the stances of the claim for both social media and news media. The developed system aims to provide a summary of the stances for the claim so that users can have a more comprehensive understanding of the information to detect potential rumors.

Keywords: Rumor detection · Claim extraction · Stance comparison

1 Introduction

Social media and online news are two major valuable sources of knowledge discovery in many applications such as business intelligence and political campaigns. They share similar contents and are interactive with each to help people have a comprehensive understanding of an event. Much of information from news articles is discussed and paraphrased on social media, and news articles may also report the trending topics discussed on the social media. The tight connection between these two information sources makes it possible to compare and analyze the public opinions about the same topic [8]. However, one major difference between social media and online news is the quality of the information. The ability that everyone can publish any information online at any time can be a

© Springer Nature Switzerland AG 2018
Y.-H. Tseng et al. (Eds.): AIRS 2018, LNCS 11292, pp. 29–35, 2018.
https://doi.org/10.1007/978-3-030-03520-4_3

double-sided sword. Although social media can facilitate healthy communication among online users, it can not make sure that all the posted information is truthful. As a result, the information from social media could be gossip, rumor or simply disinformation. To prevent the rumor spreading and enable accurate data analysis on social media, it would be necessary to develop tools to help users detect potential rumors.

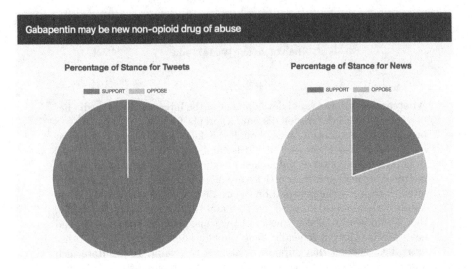

Fig. 1. Comparing the stance distributions of discussions from Twitter (left) and news articles (right)

Most existing studies tackled the problem of rumor detection as a binary classification problem and aimed to automatically identify whether a claim is a rumor based on various features, such as its content, context and spreading pattern [2,3,6,7,9]. However, these studies are not sufficient to address the problem of rumor detection. First, the detection accuracy is far from being perfect, so users can not fully trust the detection results. Even if the claim is determined as a rumor or non-rumor, users might still want to browse all the relevant information to make their own judgments or understand the detection results by themselves. Second, all these detection methods require training data, but it is labor-intensive and time-consuming to create the training set since it requires manual examination of all the relevant information related to the claim. When checking whether a claim is a rumor, users often need first to gather relevant information about the claim and go through them to analyze whether the claim is rumor or not. They may check the relevant information on social media to see how the information has been spread and what the users say about it. In the meantime, they may also check the relevant information on other information sources such as news articles and trustworthy web pages to gather more facts about the claim. As far as we know, there are no tools that can facilitate this process. Clearly, to reduce the time and effort spent on confirming a claim, it would be necessary to develop an information system that can allow users or assessors to navigate through the relevant information about the claim easily.

In this demo paper, we describe our efforts on developing the *StanceComp* system, which aggregates and compares the stances of a claim from different sources. It allows users to access the information in a top-down fashion. The intuition is that truthful information tends to have a similar distribution of the stances across different information sources. In our system, we focus on comparing the stances of the discussions from Twitter and news articles. Given a trending hashtag on Twitter, the system would first identify a few claims that are related to the hashtag. For each claim, the system would provide a quick summary of the discussions about the claim and compare the stance distributions of the discussions from tweets and news articles. If the stance distributions of the discussions from Twitter and news articles are consistent, it indicates the claim might not be a rumor. On the contrary, if the stances are completely opposite, it would indicate the possibility of the rumor. As an example, let us consider a claim - "Gabapentin may be the new non-opioid drug of abuse". Figure 1 shows the results generated by our system. The left pie graph shows the stance distribution based on the information collected on Twitter, and the right one shows the results based on news articles. It is clear that the stances are inconsistent from both sources, which indicates that the claim could be a rumor. Also, this claim is indeed a rumor as confirmed by Snopes[1].

In summary, the *StanceComp* system provides a top-down strategy for users to navigate through the information related to a trending hashtag on Twitter. For each hashtag, the users would be able to select one of the automatically generated claims and compare the stance distributions of the discussions from tweets and news articles. For each information source, the users can further drill down to browse relevant information under each stance. Finally, the developed system is complementary to the existing studies on rumor detection since it can be used to help users better understand the detection results and facilitate the creation of training data.

2   System Overview

The developed *StanceComp* system aims to provide a top-down solution for users to detect rumors. Given a hashtag, the system first crawls relevant information from Twitter and automatically extract a list of claims that are related to the hashtag. For example, given a hashtag "#germanwings-crash", the extracted claims include: (1) "A320 disappeared at 09:30 UTC from rador", (2) "Germanwings Airbus A320 crashes in French Alps near Digne", (3) "Co-Pilot Was MUSLIM CONVERT Hero of Islamic State", and (4) "Germanwings co-pilot had serious depressive episode". For each extracted claim, the system would gather relevant information from both tweets and news articles, conduct stance classifications on the gathered information, and generate a summary of the stance

---

[1] https://www.snopes.com/fact-check/gabapentin-newest-prescription-drug-killer/.

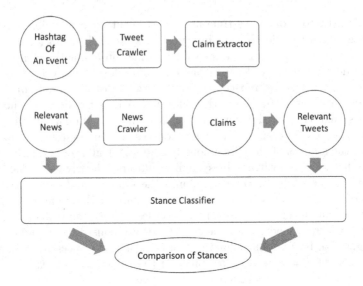

Fig. 2. The workflow of the *StanceComp* system. Data are represented as circles, and each system component is represented as a square.

distributions for each information source. Figure 2 summarizes the workflow of the *StanceComp* system, and we now provide more details for each component.

- **Tweet Crawler:** Given a hashtag, the straightforward way is to use Twitter API to crawl all the tweets containing the hashtag. However, since an event could be described by multiple hashtags, we use the co-occurrences of hashtags in the tweets to detect related hashtags in order to collect more relevant tweets. Moreover, we also used GetOldTweets-python[2] to crawl older relevant tweets.
- **Claim Extractor:** Since there could be multiple claims related to an event, this component aims to automatically extract all the claims from the crawled tweets. The first step is to generate possible claims. It is easy and efficient for us to only consider the simple sentence, which consists of the subject, verb, and object, as the claim. To achieve this goal, we first use a dependency parser for English tweets [4] to parse the tweets and select sentences whose root is a verb. Then the previous and the subsequent nouns of the root verb term on which these two are dependent could be extracted from the tweets as the subject and the object, respectively. However, simply using the single term to represent the subject or the object can not fully construct the claim embedded in the tweet for two reasons. On the one hand, the subject and the object may contain multiple terms. On the other hand, additional components should be considered to fully represent the whole meaning of the sentence, such as object complement. Therefore, we identified seven possible components based on the experiments including noun, pronoun, verb, adjective, adverb, preposition

---

[2] https://github.com/Jefferson-Henrique/GetOldTweets-python.

conjunction, and coordinating conjunction. We utilized the recursive method to explore all possible dependencies between them to find the intact subject, object, and the other necessary components to form a simple sentence. After generating all possible claims, the second step is to merge similar claims. In particular, we use the Sent2Vec pre-trained model [5] to map claims to vector space and use the DBSCAN (i.e., Density-based spatial clustering of applications with noise) method implemented in scikit-learn library[3] to find and merge similar claims. For each claim cluster, a representative claim will be selected based on the popularity, which is computed based on the number of retweets, comments, etc.

- **News Crawler:** Given a claim, this component crawls relevant news articles returned by Google. To achieve this goal, we first use representative claims to generate queries and then search them on Google to get the top 20 news articles from it. Since each news article could be long, we only used the Google snippets as claims for the news articles.

- **Stance Classifier:** For each crawled relevant information, either a tweet or a snippet of a news article, we could classify it based on its stances using the method proposed by Augenstein et al. [1]. This method leverages two neural networks based on conditional encoding to classify stances when no targets of stances shown in tweets. One long-short term memory(LSTM) network is used to encode the target and the other LSTM is used to encode the tweet with the initial state based on the previous target information. After that, the system summarizes the stance distribution for each source by computing percentages of different stances, and compare them in the pie graphs.

Fig. 3. Screenshot of claims

---

[3] http://scikit-learn.org/.

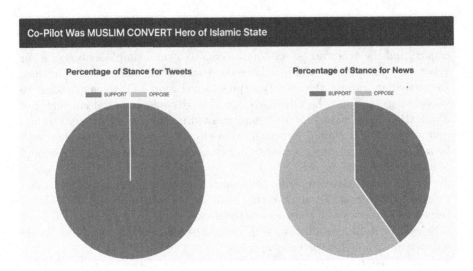

Fig. 4. Screenshot of stance distributions of the discussions from the tweets and news article

3 Conclusions and Future Work

The developed *StanceComp* is expected to facilitate the process of checking/verifying rumors. Its top-down design aims to allow users to gain a high-level view of the stance distributions before drilling down to the details. The system can be considered as a complementary effort to existing studies on rumor detection since it could automatically identify claims based on a hashtag and could facilitate the process of checking the claims.

There are many interesting future directions that we plan to work on. First, the system allows users to provide feedback on the relevance status or stance classification for each result. It would be interesting to study how to utilize the feedback in real time to refine the underlying methods. Second, the information from news articles could have a potential delay compared with the information from the tweets. We plan to study the impact of such delay on the utility of our system and explore new strategies to solve the problem. Finally, it would be interesting to conduct a more quantitative evaluation on the effectiveness of the system.

Acknowledgements. This research was supported by the University of Delaware Cybersecurity Initiative (UD CSI) Research Grant.

References

1. Augenstein, I., Rocktäschel, T., Vlachos, A., Bontcheva, K.: Stance detection with bidirectional conditional encoding. In: Proceedings of the 2016 Conference on Empirical Methods in Natural Language Processing (2016)
2. Hamidian, S., Diab, M.T.: Rumor detection and classification for twitter data. In: Proceedings of the Fifth International Conference on Social Media Technologies, Communication, and Informatics (SOTICS), pp. 71–77 (2015)
3. Hassan, N., et al.: ClaimBuster: the first-ever end-to-end fact-checking system. Proc. Very Large Data Base Endow. **10**, 1945–1948 (2017)
4. Kong, L., Schneider, N., Swayamdipta, S., Bhatia, A., Dyer, C., Smith, N.A.: A dependency parser for tweets. In: In Proceedings of the 2014 Conference on Empirical Methods in Natural Language Processing (EMNLP) (2014)
5. Pagliardini, M., Gupta, P., Jaggi, M.: Unsupervised learning of sentence embeddings using compositional n-gram features. In: NAACL 2018 - Conference of the North American Chapter of the Association for Computational Linguistics (2018)
6. Qazvinian, V., Rosengren, E., Radev, D.R., Mei, Q.: Rumor has it: identifying misinformation in microblogs. In: Proceedings of the Conference on Empirical Methods in Natural Language Processing. EMNLP 2011 (2011)
7. Tolosi, L., Tagarev, A., Georgiev, G.: An analysis of event-agnostic features for rumour classification in twitter. In: Tenth International AAAI Conference on Web and Social Media (2016)
8. Tsagkias, M., de Rijke, M., Weerkamp, W.: Linking online news and social media. In: Proceedings of the Fourth ACM International Conference on Web Search and Data Mining, pp. 565–574. WSDM 2011. ACM, New York (2011). https://doi.org/10.1145/1935826.1935906
9. Zhao, Z., Resnick, P., Mei, Q.: Enquiring minds: early detection of rumors in social media from enquiry posts. In: Proceedings of the 24th International Conference on World Wide Web, pp. 1395–1405. International World Wide Web Conferences Steering Committee (2015)

Personalized Social Search Based on Agglomerative Hierarchical Graph Clustering

Kenkichi Ishizuka[✉]

Dwango Co., Ltd., Kabukiza Tower., 4-12-15 Ginza, Chuo-ku,
Tokyo 104-0061, Japan
kenkichi_ishizuka@dwango.co.jp

Abstract. This paper describes a personalized social search algorithm for retrieving multimedia contents of a consumer generated media (CGM) site having a social network service (SNS). The proposed algorithm generates cluster information on users in the social network by using an agglomerative hierarchical graph clustering, and stores it to a contents database (DB). Retrieved contents are sorted by scores calculated according to similarities of cluster information between a searcher and authors of contents. This paper also describes the evaluation experiments to confirm effectiveness of the proposed algorithm by implementing the proposed algorithm in a video retrieval system of a CGM site.

Keywords: Social search · Graph clustering · Louvain method

1 Introduction

A social search which uses relationships between searchers and authors of contents on a social network is effective on a contents retrieval in SNSs [1]. Multimedia contents, such as videos, musical works, and/or live programs, aren't necessarily be explained enough by natural languages. It is considered that it is possible to enhance a retrieval precision of multimedia contents by using relationships between searchers and authors of contents on the CGM site having the social networks. Several studies on the personalized social search that stores relationships between users and documents to inverted index files of a search engine and uses them to score hitted documents are carried out [2,3]. Implementations at a low cost of these algorithms on the SNS having large scale social networks are difficult because a data size of the search engine becomes larger as the number of users and documents get larger. This paper describes a personalized social search algorithm for retrieving multimedia contents of a CGM site having a SNS. The proposed algorithm generates cluster information on users in the social network by using an agglomerative hierarchical graph clustering, and stores it to a contents DB. Retrieved contents are sorted by scores calculated according to similarities of cluster information between a searcher and authors of

© Springer Nature Switzerland AG 2018
Y.-H. Tseng et al. (Eds.): AIRS 2018, LNCS 11292, pp. 36–42, 2018.
https://doi.org/10.1007/978-3-030-03520-4_4

contents. The personalized social search is able to be implemented by using the proposed algorithm with small data usage even if there are a lot of users in the social network because it doesn't require direct relational information between a searcher and authors of contents. Evaluation experiments are conducted to confirm effectiveness of the proposed algorithm by implementing the proposed algorithm as a video retrieval system of NicoNico Douga which is CGM site in Japan, and by checking click through rate of video retrieval by users. In this paper, a follow relationship and a watch relationship between a searcher and authors of contents are used to retrieve multimedia contents. The watch relationship is generated by mapping user's recently video watching histories in the CGM site to a relationship between audiences and authors of videos.

The louvain method [4] which is an agglomerative hierarchical graph clustering algorithms is applied to calculate cluster information on users in CGM site, because it is able to extract hierarchical structure of communities from large networks in a short time. Retrieved contents are reranked according to scores which are calculated considering hierarchical structures of communities of social networks extracted by the louvain method.

2   System Structure

2.1   Outline of the Proposed Algorithm

A retrieval system based on the proposed algorithms consists of three sections, a cluster information generation section and a contents retrieval section and a contents data import section as shown in Fig. 1. The cluster information generation section (1) generates the cluster information of users by applying the louvain method to the follow relationship and the watch relationship of users, and (2) stores it to cluster information DB. The contents data import section (3) inputs information on newly uploaded content, (4) obtains the cluster information of the author of the content from the cluster information DB, and (5) stores the content with the cluster information to the content DB. The contents retrieval section (6) inputs requests of contents retrieval by a searcher, (7) obtains the cluster information of searcher from the cluster information DB, and (8) retrieves contents DB by a query inputted by the searcher. Retrieved contents are sorted based on scores calculated according to similarities of the cluster information between the searcher and authors of contents.

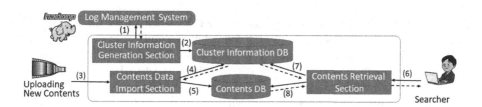

Fig. 1. Flow of the proposed algorithm

2.2   Cluster Information

The cluster information attached to users and contents is explained in this section. Hierarchical cluster structures are derived from the follow relationship and the watch relationship by using the louvain method. Because the cluster information of the follow relationship and the watch relationship are represented as a same format, this section describes the representation of the cluster information of the follow relationship.

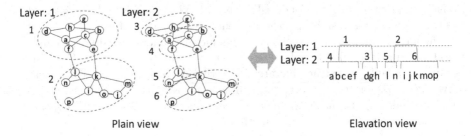

Plain view   Elavation view

Fig. 2. A plain view and an elevation view of the hierarchical cluster structure

Representation of Cluster Information. A plain view and an elevation view of each layer of the hierarchical cluster structure derived from a network of a follow relation consisting of users $a, b, ..., p$ are shown in Fig. 2. The numbers of $1, 2, ..., 6$ attached to each community in Fig. 2 are the community IDs. As can be seen from the Fig. 2, rough communities are formed in an upper layer and detailed communities are formed in a lower layer. Let N is defined as the number of layers of the hierarchical cluster structure derived by using the louvain method, and $C_{(x,i)}$ is defined as the community IDs in the i-th layer that an user x belongs to. The follow cluster information F_x of the user x is defined as:

$$F_x = [C_{(x,1)}, C_{(x,2)}, ..., C_{(x,N)}] \tag{1}$$

The user a belongs to the community 1 on the first layer, and belongs to the community 4 on the second layer in the example of Fig. 2. Therefore, the follow cluster information of the user a is $F_a = [1, 4]$. A cluster information of an author of contents is given to each content in the contents DB.

Data Size of the Cluster Information. Let N is defined as the number of the layer of the cluster information, D is defined as the number of contents in the contents DB, and L[Byte] is length of the community ID. Amount of increase of data size by appending cluster information to the contents DB is calculated by $L \times N \times D$. Since the number of the layer of the cluster structure derived from a network is not much difference by increasing of the number of nodes and edges of the network, a data size of the contents DB increases little as the number of user in the SNS is getting larger.

2.3   Scoring of Contents

Contents are scored by following two step procedures in the present system to shorten an amount of time to score the contents.

The first step: contents are scored by using freshness of them.

The second step: contents are scored by using the cluster information of a searcher and the cluster information of authors of contents.

Personalized Sort by Using the Cluster Information. When the present system receives request of contents retrieval, it obtains the cluster information of the searcher from the cluster information DB. Contents are scored by using the cluster information of a searcher and the cluster information of authors of the contents, those are reranked so that those will be interested in by the searcher are in higher place. Scores of contents are calculated by the weighted sum of six parameters. The scores of contents are defined as:

$$Score = w_0p_0 + w_1p_1 + w_2p_2 + w_3p_3 + w_4p_4 + w_5p_5 \tag{2}$$

where p_0 is a relevance of words between a search query and a description of the content, p_1 is a view count of the content, p_2 is a freshness of the content, p_3 is a freshness of a comment written to the content, p_4 is a follow cluster information similarity, and p_5 is a watch cluster information similarity. Let t_u is a content uploaded time, t_r is a content retrieved time, and Y is seconds in a year. p_2 is defined as $p_2=(t_u - t_r+Y)/Y$. If p_2 is less than 0, p_2 is reset to 0. p_3 is calculated in the same way as p_2.

Calculation of the Cluster Information Similarity. The cluster information similarity is calculated according to a similarity of the cluster information between the searcher and the author of the contents. Therefore the follow cluster information similarity and the watch cluster information similarity are calculated by a same procedure, this section describes a calculation of the follow cluster information similarity. Let the follow cluster information of the searcher A is defined as $F_A = [C_{(A,1)}, C_{(A,2)}, ..., C_{(A,N)}]$ and the follow cluster information of the author B is defined as $F_B = [C_{(B,1)}, C_{(B,2)}, ..., C_{(B,N)}]$. The follow cluster information similarity between A and B is defined as:

$$Sim(F_A, F_B) = \frac{1}{N}\sum_{i=1}^{N} f(C_{(A,i)}, C_{(B,i)}) \tag{3}$$

$$f(x, y) = \begin{cases} 1 \ (x = y) \\ 0 \ (x \neq y) \end{cases} \tag{4}$$

The follow cluster information similarity of user a and user b is 1, that of user a and d is 0.5, and that of user a and l is 0 in an example of Fig. 2.

3   Evaluation Experiment

We implement the proposed algorithm as a video retrieval function of NicoNico Douga [5] which is a CGM site in Japan and let the users in the CGM site use the video retrieval function in order to evaluate the proposed algorithm. The CGM site has a function of the SNS, and the users in the CGM site are able to follow any user they like. The users in the CGM site are able to receive an alert when the followed user uploads a new video. The follow cluster information is generated by applying the louvain method to this user's follow relationship in the CGM site. Video watching histories which the users watch to the end are extracted from all of the video watching histories on seven days. The watch relationships are generated by mapping video watching histories by users in the CGM site to relationships between audiences and authors of videos, and the watch cluster information is generated by applying louvain method to them. The sizes of the follow relationship and the watch relationship on April 2, 2018 are shown in Table 1. The cluster information of six layers is obtained by applying the louvain method to the relationships of Table 1. The retrieval system of the CGM site has approximately 15 million videos as of April 2, 2018. The parameter R which is the number of contents to be sorted by scores calculated by Eq. (2) is adjusted so that the retrieval system in the CGM site is able to response to retrieval request to less than 1 second, and is set to 50,000.

Table 1. The size of networks

The kind of networks	The number of nodes	The number of edges
The follow relationship	9,194,596	109,471,657
The watch relationship	3,047,813	48,378,044

3.1   Experimental Procedure

We evaluate whether the cluster information similarity is effective to search contents which a searcher is interested in or not by checking varies of click through rates of video retrieval by users when values of weight parameters w_4 and w_5 in Eq. (2) set to from 0.1 to 0.3. Since a purpose of this experiment is an evaluation of an effectiveness of the cluster information similarity, the values of $[w_0, w_1, w_2, w_3]$ are fixed to $[0.1, 0.8, 0.5, 0.5]$. The searchers in the CGM site are being introduced that the video retrieval function is the personalized search by using searcher's video watching histories. The patterns that w_4 and w_5 are set to 0.0 are not tried, because those patterns doesn't reflect searcher's video watching histories at all and seems to be broken.

3.2   Experimental Result

The click through rate when the weight parameter w_4 of the follow cluster information similarity and the weight parameter w_5 of the watch cluster information similarity are varied from 0.1 to 0.3 are shown in Table 2. The numbers in parentheses in Table 2 are the numbers of impressions of retrieval results to user. It is found that click through rates of the weight parameter patterns of $(0.1, 0.2)$ $(0.3, 0.2)$ $(0.2, 0.3)$ are greater than that of $(0.1, 0.1)$. The ryan's multiple comparison method is used to statistically test of differences of the click through rates of the weight parameter patterns. The null hypotheses are rejected. The click through rates of the weight parameter patterns whose weight values of the cluster information similarity are larger than 0.1 are greater than that of the weight parameter pattern whose weight values of the cluster information similarity are 0.1. It is found that the cluster information similarity is effective to search contents which a searcher is interested in.

Table 2. Click through rates of retrieval results

		A weight value of the watch cluster information similarity w_5		
		0.1	0.2	0.3
A weight value of the follow cluster information similarity w_4	0.1	41.6% (36,808)	43.2%** (37,978)	42.0%(38,327)
	0.2	42.1% (38,856)	42.1% (38,999)	43.3%** (38,433)
	0.3	42.4% (38,093)	43.4%** (37,876)	42.7%(37,728)

4 Conclusion

This paper described a personalized social search algorithm for retrieving multimedia contents of a CGM site having a SNS. The proposed algorithm sorts retrieved contents by using scores according to the cluster information of users generated by applying the louvain method to the social networks in the CGM site. The personalized social search is able to be implemented by using the proposed algorithm with small data usage. The evaluation experiments were carried out to confirm effectiveness of the proposed algorithm by implementing the proposed algorithm as a video retrieval system of a CGM site. It is found that the cluster information is effective to search contents which a searcher is interested in. A comparison with other clustering algorithms and a parameter tuning of the ranking score function by using a learning-to-rank method are future works.

References

1. Carmel, D., et al.: Personalized social search based on the user's social network. In: Proceedings of the 18th ACM Conference on Information and Knowledge Management, Hong Kong, China, pp. 1227–1236 (2009)
2. Curtiss, M., et al.: Unicorn: a system for searching the social graph. In: Proceedings of the 39th International Conference on Very Large Data Bases, Riva del Garda, Trento, Italy, pp. 1150–1161 (2013)

3. Horowitz, D., Kamvar, S.D.: The anatomy of alarge-scale scocial search engine. In: Proceedings of the 19th International Conference on World Wide Web, Raleigh, NC, USA, pp. 431–440 (2010)
4. Blondel, V.D., Guillaume, J.-L., Lambiotte, R., Lefebvre, E.: Fast unfolding of communities in large networks. J. Stat. Mech. Theor. Exp. (2008). International School for Advanced Studies and IOP Publishing
5. Nico Nico Douga. http://www.nicovideo.jp. Accessed 12 Sept 2018

Search

Improving Session Search Performance with a Multi-MDP Model

Jia Chen[✉], Yiqun Liu, Cheng Luo, Jiaxin Mao, Min Zhang,
and Shaoping Ma

Department of Computer Science and Technology, Institute for Artificial Intelligence,
Beijing National Research Center for Information Science and Technology,
Tsinghua University, Beijing 100084, China
chenjia0831@gmail.com, yiqunliu@tsinghua.edu.cn
http://www.thuir.cn

Abstract. To fulfill some sophisticated information needs in Web search, users may submit multiple queries in a search session. Session search aims to provide an optimized document rank by utilizing query log as well as user interaction behaviors within a search session. Although a number of solutions were proposed to solve the session search problem, most of these efforts simply assume that users' search intents stay unchanged during the search process. However, most complicated search tasks involve exploratory processes where users' intents evolve while interacting with search results. The evolving process leaves the static search intent assumption unreasonable and hurts the performance of document rank. To shed light on this research question, we propose a system with multiple agents which adjusts its framework by a self-adaption mechanism. In the framework, each agent models the document ranking as a Markov Decision Process (MDP) and updates its parameters by Reinforcement Learning algorithms. Experimental results on TREC Session Track datasets (2013 & 2014) show the effectiveness of the proposed framework.

Keywords: Session search · Markov Decision Process
Reinforcement learning

1 Introduction

Session search is an unfading topic in Information Retrieval (IR) research. It refers to the retrieval and reranking of documents for a search session with multiple queries. When a user tries to complete some complicated tasks, she may need information concerning various aspects. A single query can hardly satisfy such complex information need, so more queries might be submitted later. Research

---

J. Chen—This work is supported by Natural Science Foundation of China (Grant No. 61622208, 61732008, 61532011) and National Key Basic Research Program (2015CB358700).

Y.-H. Tseng et al. (Eds.): AIRS 2018, LNCS 11292, pp. 45–59, 2018.
https://doi.org/10.1007/978-3-030-03520-4_5

frameworks designed for single-round search processes may not be expert in dealing with these scenarios. To shed light on this research question, researchers in the IR community seek to optimize users' whole-session experiences with context information, e.g. "short-term" query history within a search session. This is the core idea of most existing studies which try to support session search.

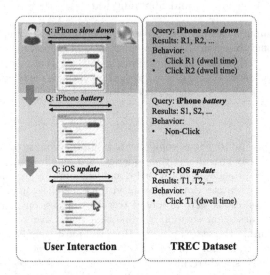

Fig. 1. An example search session which shows multi-round search interactions. (The left column illustrates the interaction process between a user and a search engine. The right column presents the information provided by TREC Session Track corpus, which is one of the most popular benchmarks in session search studies.)

Although this line of research has already gained significant success in supporting session search, we find that they may suffer from the "intent shift problem", i.e. when users are submitting queries and reading search results, their information needs keep evolving because the search can also be regarded as a learning process [7]. Figure 1 presents an example of a search session in which a user found that her iPhone became very slow and issued a query "iPhone slow down". She clicked several results and found that it may be caused by the battery problem. Then she submitted another query "iPhone battery" to the search engine and read a document which said that this issue might be solved by a future iOS update. At last, she checked the news about "iOS update". We can observe that during the example search process, the intent of the user actually shifted from "slow down" to "battery", then ended with "iOS update". This kind of "intent shift" was also observed by previous studies [12,19]. Traditional session search approaches may fail to deal with this kind of intent shift since their optimization strategies usually implicitly assume that search sessions are focusing on a specific topic/intent.

In a search session, users' interaction processes are modeled as a *trial-and-error* process, which means that users are trying to optimize their search strategies by multi-round iterations. This process can also be regarded as a gaming process in which both users and search engines only know a portion of information. Reinforcement Learning (RL) algorithms can be naturally introduced to solve these problems. For instance, [13] proposes a system called "win-win" search model which regards the search process within a session as a Partially Observable Markov Decision Process (POMDP). It assumes that the users' decision states are determined by exploration and exploitation processes of search engine users. Many existing studies share similar ideas, i.e. using the queries and results in previous rounds to improve the ranking performance of lately submitted queries.

However, most of the existing models maintain a single RL agent. The static system framework may hurt the performance of document ranking if the queries are rather dissimilar, or in other words focus on very different topics. To settle this problem, we propose a novel RL framework named Multiple Markov Decision Process (Multi-MDP) Model. The basic idea is: to improve the document rank of a certain query, we can not only learn from previously issued queries in the same session but also benefit from those semantically similar ones in other sessions. In our proposed framework, an MDP agent will be trained on a self-clustered query collection. Firstly each query will be distributed to the most appropriate cluster, then its candidate documents will be ranked by the MDP agent trained on this cluster. The system performance is validated on two popular session search benchmarks: TREC Session Track 2013 & 2014. Experimental results demonstrate that Multi-MDP model outperforms existing session search solutions on these datasets.

To summarize, the contributions of this paper are:

- We propose a novel framework based on reinforcement learning named Multi-MDP to improve performance in the context of session search.
- With Multi-MDP, we are able to capture users' evolving information needs by query clustering during the search process.
- Self-adaption techniques are employed to reduce the computational cost of multiple MDP rankers, which greatly improve the efficiency of Multi-MDP.

2 Related Work

2.1 Session Search

Search engines are trying to support complex search tasks with useful history information. There has been much work on session search [3–6,8]. These approaches can be classified into *log-based* and *content-based* methods.

Considerable work involves studying queries and sessions through session log. [16] improved ranking performances by using query chains and click-through data from logs with a learning-to-rank framework. [18] put forward an optimal rare query suggestion framework by leveraging implicit feedbacks from users in the query logs. The framework also used a random walk model to optimize query correlation. In addition, [20] targeted the identification of *cross-session* tasks by learning inter-query dependencies from user behaviors in logs with a semi-supervised clustering model. Similarly, Kotov et al. proposed methods for modeling and analyzing user search behaviors that extend over multiple search sessions [10].

Less common in session search are content-based methods which directly study the content of queries and documents. Some methods are generated with the aid of Session Tracks at TREC [1,2]. [5] greatly boosted the search accuracy by formulating structured queries using the nuggets generated from context information rather than just using the current query. Furthermore, [17] optimized the whole-session relevance by exploring the intrinsic diversity of sessions.

What sets our work apart from previous studies is: rather than looking into the relationship between query and document, we put the emphasis on the semantic similarity between queries. We detect the users' information intents by query clustering and then optimize ranking performances with Reinforcement Learning.

2.2   Reinforcement Learning in IR

In these years, Reinforcement Learning (RL) has been a promising direction in Artificial Intelligence studies. In particular, session search can be modeled as a Markov Decision Process (MDP) or Partially Observable Markov Decision Process (POMDP) based on its *trail-and-error* specialty. Many relevant problems could be solved by a family of reinforcement learning algorithms. A number of MDP/POMDP models have been raised to optimize the system performance in session search.

[6] proposed a novel query change retrieval model, named as QCM, which modeled session search as an MDP process and utilized syntactic editing changes between query change and previously retrieved documents to enhance search accuracy. Zeng and Xu's work regarded the construction of a document ranking as a sequence of decision makings and each of which corresponded to an action of selecting a document for a specific position [21].

Some other researchers exerted POMDP to construct more complicated systems [11]. [22] modeled log-based document re-ranking as a POMDP which can complete the task effectively without document contents. Yuan and Wang's work addressed online advertisement recommendation by employing POMDPs. They also discussed how to utilize the correlation of ads to improve the efficiency of the exploration and increased ads incomes. The win-win search model proposed by [13] considered session search as a dual-agent stochastic game and designed four user decision states in its POMDP.

Our approach models the re-ranking of document lists as a Markov Decision Process (MDP). The essential difference between our model and the previous one is that our framework owns a self-adaption mechanism. Specifically, all of the aforementioned systems mainly focused on optimizing search results with a static user intent while ours starts from the semantic information embedded in queries. It provides users with better search experiences when their information needs are dynamically changing.

3   Multi-MDP Ranking Model Framework

3.1   Main Framework

We introduce a Multi-MDP ranking model to improve the ranking performance in session search. The model owns multiple agents and maintains each agent by training queries with similar semantic features with reinforcement learning algorithms.

To facilitate the presentation of our work, we first briefly introduce the kernel modules of the model. Figure 2 outlines two core parts of the Multi-MDP Ranking Model.

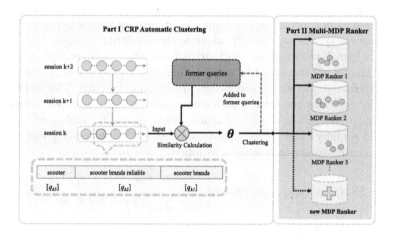

Fig. 2. The Framework of multi-MDP Ranking Model

Part I is an automatic clustering module distributing training queries into clusters according to their semantic features. Here we detect the evolving intents of search users by clustering a sequence of queries on the semantic level. For a specific query, it might be aggregated to an existing cluster or a newly created one. In this process, a hyper-parameter θ is introduced as the threshold of the minimum similarity between two queries within the same agent. We calculate the value of the θ by implementing a greedy search method on the training set. More details of this method will be described in Sect. 3.2. Then we model query clustering as a **Chinese Restaurant Process (CRP)**. In probability theory, the Chinese Restaurant Process provides a generative description for the Dirichlet Process [15]. It is a discrete-time stochastic process analogous to seating customers at tables in a Chinese restaurant. Note that CRP is introduced in the prevention of explosive growth in the number of ranking clusters in part II.

Part II is a series of ranking agents, each of which models document ranking as an MDP and trains parameters by REINFORCE, a widely-used Policy Gradient algorithm in reinforcement learning. As described in part I, the increase of the agent totality will be under the control of CRP.

System parameters are learned in the training phase. For a query in the test set, the system will choose the most appropriate cluster for it (even the cluster does exist yet and should be created). Then the query will be processed by the MDP ranker which is trained on that cluster. Note that the proposed framework can dynamically detect the instant information needs of users in multiple search rounds (Part I) as well as optimize the document list returned for a specific query at each rank (Part II).

3.2   Chinese Restaurant Process Based Query Clustering

We model the clustering process as a Chinese Restaurant Process (CRP). The reason we choose CRP rather than some other classical clustering methods is that it can dynamically control the total number of clusters with a hyper-parameter ξ. For a new query, the system will calculate the similarity between it and existing clusters first. If the maximal similarity is lower than a pre-set threshold, then the query should not be owned by any existing clusters. So it will be distributed to a newly created one with a probability determined by CRP. This mechanism efficaciously restrains the total number of rankers within an upper bound and avoids a huge occupation of system space. For a query Q_i (analogous to a customer) and a query cluster (compared to a table), the probability distribution of seating one customer at a table is as follows:

$$Pr(Q_i = r | Q_1, ..., Q_{i-1}) = \begin{cases} \frac{\xi + |B|\alpha}{\xi + i - 1} & if \ r \ \in \ new \ ranker, \\ \frac{|b| - \alpha}{\xi + i - 1} & if \ r = b, b \in B; \end{cases} \tag{1}$$

where $|B|$ denotes the total number of existing clusters, b is one of the non-empty clusters and $|b|$ denotes the capacity of it. Specifically, we only need to control the probability of creating a new cluster so only the first formula will be utilized. Thus there is no need to discuss the value of $|b|$ here. If we let $\alpha = 0$, then a single parameter ξ needs to be determined. This can be solved by the expected number of rankers $E(R)$ as shown in Eq. 2:

$$E(R) = \sum_{k=1}^{n} \frac{\xi}{\xi + k - 1} = \xi \cdot (\Psi(\xi + n) - \Psi(\xi)), \tag{2}$$

In this equation, n refers to the number of queries. ξ is the hyper-parameter while $\Psi(\xi)$ denotes the *digamma* function. To hold system robustness, we set the expected number of rankers $E(R)$ as twice of topic scopes in training set (this can be adjusted according to system condition). By solving the equation, the value ξ can be determined and will be introduced as a parameter in query clustering. In this way, the total number of rankers will converge to an invariant value rather than increase endlessly.

Given that query clustering has a direct impact on the system performance, we should try to find a potentially reasonable threshold, e.g. θ, which can discriminate a query from other dissimilar ones. If the similarity of the two queries is higher than the threshold then we should aggregate them in the same cluster. One of the options is to estimate θ based on the topic labels provided by TREC Session Track training set. A greedy algorithm searching for the value θ follows two heuristic rules:

- if the maximum similarity of the current query q and a previous query $\hat{q}$ is greater than the threshold, then $label_q = label_{\hat{q}}$;
- if the maximum similarity of the current query q and a previous query $\hat{q}$ is lower than the threshold, then $label_q$ is assigned to a new label.

In the training phase, the clustering process is performed as mentioned above. However in testing phase, a query is always distributed to the cluster that is the most similar to it. By comparing the coherence with ground truth labels for various searching values of θ, we choose the one of highest coherence. The ground truth labels are analyzed from the log provided by TREC which classifies each query to a corresponding topic.

3.3 MDP Ranking Agent

Having clustered each query into a specific cluster, the system should optimize document rankings for queries within a cluster. The document ranking process for each query is formalized as a Markov Decision Process (MDP) [21], where document ranking is considered as a sequence of decision makings, each of which can be defined as selecting one appropriate document at a rank in vertical position. As shown in Fig. 3, the proposed MDP ranking process is represented as a tuple $\{S, A, T, R, \pi\}$, respectively denoting system states, actions, transition function, reward and the decision policy. Detailed expositions are described as:

States S is a set of different states that describe the system environment. To arrange a rank for a document set, the agent should make a sequence of decisions. One decision can be made according to the position and the candidate document set the agent can choose from. Thus, the state at the time t is designed as $[t, X_t]$, where X_t represents the remaining candidate document set.

Actions A refers to a set of actions that the agent could take at each step. In our model, the agent should decide which document to be selected at the current position. The action set for the state s_t is denoted by $A(s_t)$. At the time t, $a_t \in A(s_t)$ selects a document d_i at the ranking position $t + 1$, where i is the index of the chosen document.

Transition Function $T(S, A)$ is a function that maps a state s_t into a new state s_{t+1} in response to the action at time t. When the agent chooses a document d_i from the candidate set and ranks it at the t-th position, the document is removed from the candidate set in case of duplication. Equation 3 is the formulation of the transition function.

$$s_{t+1} = T([t, X_t], a_t) = [t + 1, X_t \backslash d_i], \tag{3}$$

Reward $R(S, A)$ is the global reward of a rank, which directly represents the gains of users. In this paper, the reward is regarded as a sum of discounted gain at each position from 1 to N. It is defined in a manner similar to DCG (a widely-used metric) to directly optimize the evaluation measures and leverages the performances at each rank. Unable to estimate accurate user gains from the insufficient interaction data, we design a reward function in direct promotion of IR metrics.:

$$R(s_t, a_t) = \begin{cases} 2^{L(d_i)} - 1 & if\ t = 0, \\ \frac{2^{L(d_i)}-1}{log_2(t+1)} & if\ t \neq 0; \end{cases} \tag{4}$$

where $L(d_i)$ is the relevance label of the selected document d_i. The reward at each position will be accumulated at the end to represent the global reward of a rank, which provides references for an agent to update parameters through RL algorithms.

Policy $\pi(a|s) : A \times S \rightarrow (0,1)$ is a probabilisitic distribution over the possible actions for an agent at the state s_t. Generally, the policy of one agent represents the probabilities of selecting one specific document for a position. This is formalized as Eq. 5:

$$\pi(a_t|s_t; w) = softmax(w^T V(d_i)), \tag{5}$$

where $V(d_i)$ denotes the vector representation of d_i, $w \in \mathbb{R}^k$ represents the parameters of each ranking agent with the same dimension k as the query/document representations.

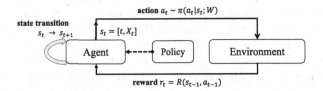

Fig. 3. State transition in each MDP ranking agent

Given a training query q, a set of h candidate documents D, the relevance labels for each document L, the initial system state can be denoted as $s_0 = [0, X]$. At each time $t = 0, ..., h-1$, the agent observes the environment and determines the belief of its state s_t. According to the state and the system policy π, an agent carries out an action a_t and ranks a document at a position. The agent will immediately receive a reward $r_t = R(s_t, a_t)$ on the basis of the relevance label. Then the system transfers to the next state s_{t+1} and tries to take next action a_{t+1}. The process is repeated until all the h documents are selected. We adopt the Monte Carlo Sampling method to generate several ranks and calculate the reward of these ranks. Of all the rank with the highest reward will be recorded for system updates.

In the testing phase, no label is available. An agent will generate a rank according to its learned policy. For each position, the agent will choose a document with maximal probability. The probability is calculated by the product of w and document representation. In order to learn the model parameters w, we use REINFORCE algorithm to update them. According to the reward defined above, the long-term gain of a sampled ranking list μ can be defined as:

$$G(\mu) = \sum_{k=1}^{M} \gamma^{k-1} r_k, \tag{6}$$

where μ is the ranking list sampled from a stochastic ranking process, on the basis of Monte Carlo sampling. Note that if $\gamma = 1$, the formula is equivalent to the definition of DCG. According to REINFORCE algorithm, the gradient $\nabla_w J(w)$ can be calculated as follows:

$$\nabla_w J(w) = \gamma^t G_t \nabla_w log \pi_w(a_t|s_t; w), \tag{7}$$

where $J(w)$ is the expected long-term return of the document ranking for a query. Our goal is to optimize this value, enabling the sampled rank to approximate the ideal rank. By comparing the reward received by the agent from different ranks, the agent chooses the one with the highest reward and stores the rank with its reward as an episode into its memory. Every time after a batch of training queries has been processed, the agent updates its parameters by the episodes stored in the memory through Eq. 7 and clears the memory then.

4   Experiments

4.1   Dataset

To validate the effectiveness of our proposed Multi-MDP framework, we conducted our experiments on TREC Session Tracks 2013–2014 datasets [1,2]. TREC Session Tracks were one of the core tasks of Text REtrieval Conference[1].

As shown in Fig. 1, in Session Tracks, the participants were given a collection of search sessions. For each session, the participants were provided with real users' interactions in multiple search rounds except for the last one. The goal is to optimize document rank for the last query with the context information in previous search rounds.

Table 1 provides an overview of the two datasets. For each query, we used the top 100 documents through batch query service as the candidate set on the ClueWeb12 B13 dataset. In addition, queries whose number of documents returned by batch service are less than 100 were filtered. All duplicated documents were removed as well.

---

[1] http://trec.nist.gov.

Table 1. Dataset statistics for TREC session track 2013 and 2014

Data	2013	2014
#topic number	61	60
#session number	133	1,257
#query number	453	3,213
#unduplicated documents	26,664	107,564

4.2   Query Clustering

The clustering process is based on semantic features. For a pair $<q_1, q_2>$, the similarity can be calculated as follows:

1. The textual similarity of q_1 and q_2 (QT);
2. The proportion of overlapped documents in the top 10 retrieved documents of q_1 and q_2 (DO);
3. The average textual similarity of top-10 retrieved documents of q_1 and q_2 (DT);

These three measurements are respectively denoted as QT, DO and DT in the remaining parts of this paper. Note that for QT and DT, textual similarity refers to the cosine similarity of two semantic representations. DT can be formulated as:

$$sim^{DT}(q_1, q_2) = \frac{1}{100} \sum_{i=1}^{10} \sum_{j=1}^{10} cos(V_{d_{1i}}, V_{d_{2j}}) \tag{8}$$

where d_{1i} refers to the document at the i-th position of q_1 and d_{2j} refers to the document at the j-th position of q_2. V denotes vector representation.

To better capture the implicit information intents embedded in queries, the similarity is calculated based on the *distributed representation* of textual contents, i.e. Another option here is to use the *one-hot* representation of tokens. We choose distributed presentations since they are more likely to capture the semantic level similarity while one-hot representations tend to capture exact matchings of textual units.

According to the three definitions of query similarity and two methods of generating document vector representations (this will be described in Sect. 4.3), five combinatorial methods can be applied in the traversal process (the result of top 10 document overlapping ratio is equivalent for max-pooling or average-pooling representations, hence a repeated method is removed). The consistency was measured by Cohen's Kappa. Table 2 shows the best thresholds θ found for each method/dataset. Note that the scale of 2014 dataset is much larger, the clustering consistency results are much lower compared to those of the 2013 dataset. Generally, the similarity between MAX representations is higher so we can also find a relatively higher value of θ in MAX representations.

Table 2. Best threshold of each method.

Method	Best θ	Kappa	Method	Best θ	Kappa
MAX_QT (2013)	0.95	0.454	MAX_QT (2014)	0.96	0.052
AVG_QT (2013)	0.75	0.335	AVG_QT (2014)	0.77	0.054
DO (2013)	0.60	0.497	DO (2014)	0.90	0.040
MAX_DT (2013)	0.99	0.289	MAX_DT (2014)	0.99	0.032
AVG_DT (2013)	0.99	0.428	AVG_DT (2014)	0.99	0.044

4.3   Evaluation Settings

We followed the Session Tracks' settings and used the first $N - 1$ queries in each session for training. The Test set is consist of the last queries. For both Session Track 2013 & 2014 dataset, the ratio of the training set and test set approximates 4:1. Queries and documents actually contain multiple words. We tried two methods of generating the query-level/document-level representations: max-pooling (MAX) and average-pooling (AVG) of word vectors. All web pages in Session Track 2013 & 2014 dataset were collected as the corpus. The detailed process is described as follows: We first fed the corpus to a word representation model, named as *Global Vectors for Word Representation* (GloVe) [14], to get each word's representation. Long-tailed words were removed and only those with top 100K frequency remain.

To validate the system performance, we compared our model with several baseline systems listed as following:

- *Win-Win*: A framework which models the session search process as Partially Observable Markov Decision Process (POMDP) [13];
- *Lemur*: We submitted each query to Lemur Indri Search Engine[2] through online batch service provided by ClueWeb12 and calculate evaluation metrics of the document lists returned from it;
- *TREC median*: The median TREC system reported by TREC Session Track;
- *TREC best*: The winning run in TREC Session Tracks;
- *QCM SAT*: A variation of QCM proposed by [6];
- *Rocchio*: A session search model which is proposed by [9].

We use nDCG and MAP to evaluate different runs. For nDCG, we choose several different cutoffs to check the performance of our model at different ranks.

4.4   Experimental Results

For both TREC 2013 and 2014 dataset, we tried our Multi-MDP Ranking Model with six types, where QT, DO and DT denote three definitions of query similarity and "MAX"/"AVG" refers to the generating methods for query/document presentation.

---

[2] http://lemurproject.org/clueweb12/.

Table 3 shows the search performance of all systems in TREC Session Tracks 2013. Since our experimental settings are exactly the same as [13], we cite the results they reported for comparison. The result shows that our model significantly outperforms other session search systems. We can see that Multi-MDP$^{MAX\text{-}QT}$ is the best among all methods. The nDCG metrics of Multi-MDP models are significantly higher than those of Lemur Search Engine at all ranks. The results validate that our model can better detect evolving intents of users and improve search performance in both whole-session level (high metric value on average) and query level (high performance at each rank).

Table 3. Search performance on TREC session tracks 2013. (The best multi-MDP model, i.e. Multi-MDP$^{MAX\text{-}QT}$, is significantly better than the strongest baseline Win-Win at $p < 0.001$ using a paired t-test.)

System	MAP	nDCG@1	nDCG@3	nDCG@6	nDCG@10
Win-Win	0.1290	–	–	–	0.2026
Lemur	0.1072	0.0592	0.0775	0.0953	0.1123
TREC best	–	–	–	–	0.1952
TREC median	–	–	–	–	0.1521
QCM SAT	0.1186	–	–	–	0.0939
Rocchio	0.1320	–	–	–	0.1060
Multi-MDP$^{MAX\text{-}QT}$	**0.2190**	**0.2561**	**0.2638**	**0.2453**	**0.2556**
Multi-MDP$^{AVG\text{-}QT}$	0.1891	0.2073	0.1956	0.2089	0.2182
Multi-MDP$^{MAX\_DO}$	0.2138	0.2561	0.2339	0.2193	0.2301
Multi-MDP$^{AVG\_DO}$	0.1342	0.1138	0.1171	0.1333	0.1424
Multi-MDP$^{MAX\_DT}$	0.1488	0.1638	0.1627	0.1571	0.1618
Multi-MDP$^{AVG\_DT}$	0.1306	0.1156	0.1075	0.1170	0.1409

Table 4 shows the search performance of all systems for TREC Session Tracks 2014. On the 2014 dataset, we can only get the results of TREC Best and TREC Median [2]. Although the dataset scale is much larger, the experimental results show that the Multi-MDP Ranking Model still performs well. Of all the methods, Multi-MDP$^{MAX\text{-}QT}$ still owns the best performance, although the results are not statistically significant compared to the TREC Best result.

An interesting finding is that the systems with QT usually outperform other systems. It suggests that the query contents are more representative than the document lists or document contents. Query contents suffer from fewer noises than documents thus could express refined user intents which may improve clustering accuracy. Another finding is that the systems with DO perform better for a dataset on a larger scale. This might be caused by the smaller threshold of small dataset which owns lower discrimination between two documents.

Table 4. Search performance on TREC session tracks 2014. (The best multi-MDP model, i.e. Multi-MDP$^{MAX\text{-}QT}$, is significantly better than the baseline Lemur at $p < 0.001$ using a paired t-test.)

System	MAP	nDCG@1	nDCG@3	nDCG@6	nDCG@10
Lemur	0.1260	0.0916	0.1009	0.1139	0.1251
TREC best	–	–	–	–	0.2099
TREC median	–	–	–	–	0.1170
Multi-MDP$^{MAX\text{-}QT}$	**0.1857**	**0.2271**	**0.2169**	**0.2118**	**0.2107**
Multi-MDP$^{AVG\text{-}QT}$	0.1315	0.1194	0.1063	0.1211	0.1361
Multi-MDP$^{MAX\text{-}DO}$	0.1850	0.2047	0.2036	0.2000	0.1995
Multi-MDP$^{AVG\text{-}DO}$	0.1516	0.1573	0.1630	0.1649	0.1751
Multi-MDP$^{MAX\text{-}DT}$	0.1293	0.1358	0.1388	0.1384	0.1412
Multi-MDP$^{AVG\text{-}DT}$	0.1586	0.1413	0.1544	0.1672	0.1779

5   Conclusion

This paper presents a novel session search framework: Multi-MDP Model, which models the document reranking for each query as a Markov Decision Process (MDP) and owns multiple agents. Our model gives a new guide of constructing a self-adaptive framework to fit with the size of the dataset and dynamically detect the instant information intents of search users. The increase of agents is under the control of the Chinese Restaurant Process (CRP) to ensure a low space occupation. In each agent, an MDP-based ranking model is trained for better document ranking performances. The experiments on TREC Session Tracks 2013 and 2014 datasets show that our work is both effective and efficient for session search and outperforms the best systems in TREC.

Our work is content-based and considers the explicit feedbacks of users as the global reward for a rank. We emphasize two main stages: query clustering and model training. Some other factors like click-through data, dwell time from session logs have not been utilized in the system yet. We may have an attempt to make better use of this information to improve the system performance in future.

References

1. Carterette, B., Kanoulas, E., Hall, M., Bah, A., Clough, P.: Overview of the TREC 2013 session track. Technical report, Delaware University Newark Department of Computer and Information Sciences (2013)
2. Carterette, B., Kanoulas, E., Hall, M., Clough, P.: Overview of the TREC 2014 session track. Technical report, Delaware University Newark Department of Computer and Information Sciences (2014)
3. Carterette, B., Kanoulas, E., Yilmaz, E.: Simulating simple user behavior for system effectiveness evaluation. In: Proceedings of the 20th ACM International Conference on Information and Knowledge Management, pp. 611–620. ACM (2011)

4. Feild, H., Allan, J.: Task-aware query recommendation. In: Proceedings of the 36th International ACM SIGIR Conference on Research and Development in Information Retrieval, pp. 83–92. ACM (2013)
5. Guan, D., Yang, H., Goharian, N.: Effective structured query formulation for session search. Technical report, Georgetown Univ Washington DC Dept Of Computer Science (2012)
6. Guan, D., Zhang, S., Yang, H.: Utilizing query change for session search. In: Proceedings of the 36th International ACM SIGIR Conference on Research and Development in Information Retrieval, pp. 453–462. ACM (2013)
7. Gwizdka, J., Hansen, P., Hauff, C., He, J., Kando, N.: Search as learning (SAL) workshop 2016. In: Proceedings of the 39th International ACM SIGIR Conference on Research and Development in Information Retrieval, pp. 1249–1250. ACM (2016)
8. Jiang, J., Han, S., Wu, J., He, D.: Pitt at TREC 2011 session track. In: TREC (2011)
9. Joachims, T.: A probabilistic analysis of the Rocchio algorithm with TFIDF for text categorization. Technical report, Carnegie-Mellon University Pittsburgh PA Department of Computer Science (1996)
10. Kotov, A., Bennett, P.N., White, R.W., Dumais, S.T., Teevan, J.: Modeling and analysis of cross-session search tasks. In: Proceedings of the 34th international ACM SIGIR conference on Research and development in Information Retrieval, pp. 5–14. ACM (2011)
11. Littman, M.L., Kaelbling, L.P., Cassandra, A.R.: Planning and acting in partially observable stochastic domains. Artif. Intell. 99–134 (1995)
12. Liu, J., Belkin, N.J.: Personalizing information retrieval for multi-session tasks: the roles of task stage and task type. In: Proceedings of the 33rd International ACM SIGIR Conference on Research and Development in Information Retrieval, pp. 26–33. ACM (2010)
13. Luo, J., Zhang, S., Yang, H.: Win-win search: dual-agent stochastic game in session search. In: Proceedings of the 37th International ACM SIGIR Conference on Research and Development in Information Retrieval, pp. 587–596. ACM (2014)
14. Pennington, J., Socher, R., Manning, C.: GloVe: global vectors for word representation. In: Proceedings of the 2014 Conference on Empirical Methods in Natural Language Processing (EMNLP), pp. 1532–1543 (2014)
15. Pitman, J., et al.: Combinatorial stochastic processes. Technical report, Technical Report 621, Department of Statistics, UC Berkeley, 2002. Lecture notes for St. Flour course (2002)
16. Radlinski, F., Joachims, T.: Query chains: learning to rank from implicit feedback, pp. 239–248 (2006)
17. Raman, K., Bennett, P.N., Collins-Thompson, K.: Toward whole-session relevance: exploring intrinsic diversity in web search. In: Proceedings of the 36th International ACM SIGIR Conference on Research and Development in Information Retrieval, pp. 463–472. ACM (2013)
18. Song, Y., He, L.W.: Optimal rare query suggestion with implicit user feedback. In: International Conference on World Wide Web, WWW 2010, Raleigh, North Carolina, USA, pp. 901–910, April 2010
19. Teevan, J., Dumais, S.T., Liebling, D.J.: To personalize or not to personalize: modeling queries with variation in user intent. In: International ACM SIGIR Conference on Research and Development in Information Retrieval, pp. 163–170 (2008)

20. Wang, H., Song, Y., Chang, M.W., He, X., White, R.W., Chu, W.: Learning to extract cross-session search tasks. In: Proceedings of the 22nd International Conference on World Wide Web, pp. 1353–1364. ACM (2013)
21. Wei, Z., Xu, J., Lan, Y., Guo, J., Cheng, X.: Reinforcement learning to rank with markov decision process (2017)
22. Zhang, S., Luo, J., Yang, H.: A POMDP model for content-free document re-ranking, pp. 1139–1142 (2014)

Analysis of Relevant Text Fragments for Different Search Task Types

Atsushi Keyaki$^{(\boxtimes)}$ and Jun Miyazaki

Tokyo Institute of Technology, 2-12-1 Oookayama, Meguro-ku, Tokyo, Japan
`keyaki@lsc.cs.titech.ac.jp`, `miyazaki@cs.titech.ac.jp`

Abstract. This paper investigates the trend of relevant text fragments by task type. The search results of fine-grained information retrieval systems propose not documents but text fragments. We hypothesize that the properties of relevant text fragments depend on the task type. To reveal these properties, we evaluate a relevant text fragment to judge (1) its granularity (e.g., word, phrase, or sentence) and (2) its structural complexity. Our analysis shows that a task type based on more complex information needs has a larger granularity of relevant text fragments. On the other hand, the complexity of task type's information needs does not necessarily correlate with the structural complexity of the relevant text fragments.

Keywords: Task type · Complexity · Text fragment
Information unit · MobileClick

1 Introduction

Information needs have become more complex as information systems and the information society advance. In this situation, an information retrieval system using a single model cannot always propose effective search results. Thus, an information retrieval system is expected to selectively use multiple models to propose search results based on a certain criterion such as *task type*. Such a criterion can be based on Broder's initial classification [2], the *cognitive process* [1] in interactive information retrieval, or Li et al.'s classification based on information needs [9].

Li et al. classified information needs into 20 categories derived from Google's search logs (Table 1). They reported that some task types are based on simple information needs, while others are based on complex ones. Li et al. also argued that user's information needs for task types with simple information needs can be directly satisfied with result snippets in SERPs (namely, no need to click on any links). This is extremely beneficial as it reduces the laborious tasks of browsing web pages. On the other hand, the result snippets for task types with complex information needs can be improved. This suggests that an effective approach is to employ strategies based on task type.

© Springer Nature Switzerland AG 2018
Y.-H. Tseng et al. (Eds.): AIRS 2018, LNCS 11292, pp. 60–66, 2018.
https://doi.org/10.1007/978-3-030-03520-4_6

The NTCIR task *MobileClick* [7], which extracts relevant text fragments, information unit (*iUnit* for short), focuses on the task types of complex information needs (e.g., *Local*, *Celebrities*, *QA*) and the frequently operated task *Definition*. The formal definition of iUnit is as follows: text fragments that are relevant to a query and atomic in interpreting information. An iUnit is composed of arbitrary granular information such as a few words, phrase, or sentence. We hypothesize that an effective approach to extract iUnits depends on the task type. This study evaluates the following research questions:

RQ1 Does the complexity of iUnits differ by task type?

RQ2 Does the granularity (word, phrase, or sentence) affect the complexity of iUnits for a given task type?

RQ3 Does the structural complexity affect the complexity of iUnits for a given task type?

Table 1. Task types, information needs, and query examples

Task type	Definition	Query example
Local	Local listing (phone number)	Museum allentown
Celebrities	News or images of a celebrity	Christopher nolan
Definition	Definition of a term	Bitcoin
QA	Short answer to a question	How to cook coleslaw
Currency	Currency conversion	1 USD in GBP
Click-to-call	User-typed a phone number into search to call it.	
Images	Pictures of a person or thing	Cubs logo
Lyrics	Song lyrics	Abettes with time lyrics
Map	Map of a location (e.g., address, geographic locations)	E 83rd St, Los Angeles, CA 90001
News	Current news on a topic	Santa cruz wild fires
Product	Product information (e.g., price, vendors)	2000 gsx 750f for sale
Person	Contact information on a non-celebrity person	
Quotation	Reference of a quoted sentence/phrase	
Stock	Current stock price	Quote AKAM
Sports	Sports scores	Chicago cubs
Showtimes	Movie showtimes	The dark knight
Spelling	Correct spelling of a word	Unfortuanatly
SMS	Short (greeting) messages to send	
Translation	Translation of a word in foreign language	Dounika translation
Weather	Weather report	Weather New York

To answer the RQs, we evaluated the iUnits of different task types to judge (1) the granularity and (2) the structural complexity (whether the iUnit is expressed in the form of a simple structure *key-value*). Our analysis shows that task types with more complex information needs have iUnits with a larger granularity. On the other hand, the complexity of the task type's information needs does not correlate with the structural complexity of the iUnits. Therefore, RQ1 and RQ2 are affirmative while RQ3 is negative.

Consequently, the appropriate granularities of the iUnits by task type are revealed in fine-grained information retrieval where an information retrieval system directly satisfies the user's information needs. This finding is promising to develop an effective fined-grained information retrieval system.

2   Task Types and MobileClick

According to Li et al.'s classification [9] (Table 1), some task types such as *Weather, Stock, Showtimes,* and *Spelling* are classified as simple information needs, and these can be directly satisfied with the result snippets in SERPs. In addition, there are some task types such as *Click-to-call, SMS,* and *Translation* where a search system is used as a trigger to run another application. For these types with complex information needs, that is, Local, Celebrities, QA, and *Images,* the generated result snippet can be improved.

MobileClick tackles four task types. Three have complex information needs (i.e., Local, Celebrities, and QA excluding non-text retrieval task Images) and the other is the frequently operated task Definition. Note our previous survey on TREC Web track topics[1] indicated that the majority of web queries are classified into these four categories, but the original query log data used in [9] is unavailable.

Table 2. iUnit Examples

Task type	Query	iUnit example	# of iUnits (ratio)
Local	Museum allentown	Tel: 610-432-4200	625 (0.22)
Celebrities	Christopher nolan	Movie: 'Interstellar'	778 (0.27)
Definition	Bitcoin	Online payment system	1,165 (0.40)
QA	How to cook coleslaw	Chop the green pepper and onion	329 (0.11)

MobileClick is a kind of a fine-grained information retrieval. It aims to provide not documents but iUnits as search results. An iUnit is composed of arbitrary granular information such as a few words, phrase, or sentence. Table 2 shows examples of iUnits. We used the MobileClick-2 dataset[2] (the second round of the NTCIR MobileClick task), which contains 100 queries[3] (20 Local queries, 20 Celebrity queries, 20 QA queries, and 40 Definition queries) and 2,317 iUnits in total. These iUnits are comprehensively extracted from relevant documents manually.

---

[1] https://trec.nist.gov/.
[2] http://www.mobileclick.org/.
[3] Some previous studies such as [10,11,13] focus on automatic task type classification. However, this study just uses the task type-tagged queries.

3   Formulation of Research Questions

We hypothesized that the design strategy should depend on the task type. We estimate that the complexity of a task type differs among the MobileClick task types. Specifically, the search results (i.e., iUnits) of task types based on more complex information are more complex. We focus on the granularity and structural complexity of iUnits. Since an iUnit in MobileClick is atomic in interpreting information, the iUnit is supposed to be expressed as smaller granules such as a word or compound word if the information needs are less complex. Conversely the text fragment is supposed to be expressed as a larger granule such as a phrase or a sentence if the information needs are more complex. With respect to the structural complexity, it is presumed that less complex information is structured. For instance, an iUnit of Local is likely to be expressed in the form of a key-value because the search results of Local are listings.

Based on the discussion above, we try to address following research questions. (RQ1) Does the complexity of iUnits differ by task type? (RQ2) Does the granularity (word, phrase, or sentence) affect the complexity of iUnits for a given task type? (RQ3) Does the structural complexity affect the complexity of iUnits for a given task type?

Table 3. Anderson and Krathwohl's cognitive processes [1]

Process	Definition
Remember	Retrieving, recognizing, and recalling relevant knowledge from long-term memory
Understand	Constructing meaning from oral, written, and graphic messages through interpreting, exemplifying, classifying, summarizing, inferring, comparing, and explaining
Apply	Carrying out or using a procedure through executing or implementing
Analyze	Breaking material into constituent parts, determining how the parts relate to one another and to an overall structure or purpose through differentiating, organizing, and attributing
Evaluate	Making judgments based on criteria and standards through checking and critiquing
Create	Putting elements together to form a coherent or functional whole; reorganizing elements into a new pattern or structure through generating, planning, or producing

It should be noted that we suppose Local and Celebrities as less complex information needs, while Definition and QA are classified as more complex information needs. Anderson and Krathwohl defined six types of cognitive processes [1], i.e., *Remember, Understand, Apply, Analyze, Evaluate,* and *Create* (Table 3). We regard Local and Celebrities are classified into Understand. On the other hand, Definition is applicable to Understand and Analyze, while QA is applicable to Understand and Evaluate. Therefore, we judged that task types based on a single cognitive process Local and Celebrities are less complex than these on multiple cognitive processes Definition and QA.

4   Tagging and Analysis

In this section, we describe the tagging of granularity and structural complexity of iUnits and the analytical results.

4.1   Tagging of Granularity and Structural Complexity

We tagged the granularity and structural complexity with iUnits using crowd sourcing. An iUnit is identified as either a *key-value, word, sentence,* or *disable* (disable to identity between key-value, word, and sentence). We asked crowd workers to regard a phrase as a sentence because our preliminary investigation showed the classification between a phrase and a sentence is difficult. First, a crowd worker judged whether an iUnit is in the form of key-value. Otherwise, the crowd worker chose the most appropriate one from word, sentence, and disable. Each iUnit was evaluated by five crowd workers for reliability[4].

4.2   Analysis of Tagging Results

Figure 1 shows the tagging results. Annotators attained reliable inter-annotator agreement (Fleiss' Kappa is 0.65). An iUnit is assigned as a key-value (kv), word, sentence, or disable when three or more of the crowd workers voted for the same option (the valid threshold is 3). Other iUnits are assigned as *invalid*. As a result, no iUnit is assigned as disable and the percentile of invalid is between 1% and 4%. Note that the percentile of invalid increases to 25% when the valid threshold is set to 4. Regardless of the threshold, the proportions of other options change only slightly. Thus, we decided to set the valid threshold to 3. The results show that the trends completely differ from each other, confirming that the complexity of iUnits differs by task type (RQ1).

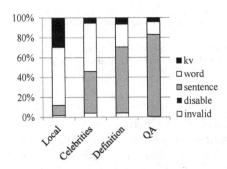

Fig. 1. Summary of the tagging results

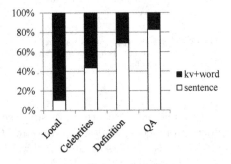

Fig. 2. Results on granularity

---

[4] http://www.lsc.cs.titech.ac.jp/keyaki/dataSet/mc2_iunitGranular_dist.tsv.

Figure 2 shows the results on granularity where the key-value and word are accumulated because most iUnits judged as key-value are reasonable to be classified as words. For simplification, disable and invalid are excluded. The results show that the granularities of Local and Celebrities are relatively small, while those of Define and QA are relatively large. Consequently, the results confirm that the granularity (word, phrase, or sentence) affects the complexity of iUnits of a task type (RQ2).

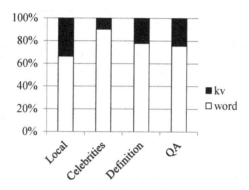

Fig. 3. Results of the structural complexity

Figure 3 illustrates the results for the structural complexity where only key-values and words are displayed. The complexity of information needs for a task type does not correlate with the structural complexity of iUnits. As expected, many iUnits of Local such as "Hours: 8:00 am – 5:00 pm" are classified as key-value. In Celebrities, many iUnits are not expressed as key-value such as "runs the production company Syncopy Inc.". This is the reason why the ratio of key-value of Celebrities is low. Some of the iUnits of Definition and QA have larger ratios than expected. For example, iUnits of Definition and QA are in the form of key-value such as "Organizations accepting bitcoin: Greenpeace, The Mozilla Foundation, and The Wikimedia Foundation" and "ingredients: cabbage head, carrot, green pepper, onion". These results indicate that the structural complexity of iUnits is independent of the task type. That is, RQ3 is not confirmed.

In summary, the relevant text fragments (iUnits) depend on the task type in terms of granularity. Hence, fine-grained task type-oriented information retrieval is expected to improve the effectiveness using the finding. It should be noted that given task type estimated with existing studies [10,11,13] appropriate granular iUnits are expected to be extracted automatically using morphological analysis or other NLP techniques which help identify the granularity of a relevant text.

5   Related Work

In this study, we consider the granularity and structural complexity to judge the complexity of a task type. Apart from our approach, a variety of metrics have been proposed to identify a task's complexity and difficulty. Examples include behavioral features [5,6], questionnaires [8], diversity of search results [4], a priori determinability [3] (whether input, process, and output can be deduced or not), and the number of subtasks and the number of operations in a query [12]. Since our approach leverages not user-related data but relevant assessment data, [4] is the most related study to ours.

6   Conclusion

We analyzed the properties of iUnits for different task types of MobileClick with an emphasis on the granularity and structural complexity. We found that granularity of iUnits depends on the task type and the complexity of the task type's information needs does not correlate with the structural complexity of the iUnits. In the future, we plan to consider granularity in generating search results for fine-grained information retrieval.

Acknowledgments. This work was partly supported by ACT-I, JST.

References

1. Anderson, L.W., et al.: A Taxonomy for Learning, Teaching, and Assessing: A Revision Of Bloom's Taxonomy of Educational Objectives. Longman, New York (2002)
2. Broder, A.: A taxonomy of web search. ACM SIGIR Forum **36**(2), 3–10 (2002)
3. Byström, K., Järvelin, K.: Task complexity affects information seeking and use. Inf. Process. Manag. **31**(2), 191–213 (1995)
4. Campbell, D.J.: Task complexity: a review and analysis. Acad. Manage. Rev. **13**(1), 40–52 (1988)
5. Hu, X., Kando, N.: Task complexity and difficulty in music information retrieval. J. Assoc. Inf. Sci. Technol. **68**(7), 1711–1723 (2017)
6. Arguello, Jaime: Predicting search task difficulty. In: de Rijke, M., et al. (eds.) ECIR 2014. LNCS, vol. 8416, pp. 88–99. Springer, Cham (2014). https://doi.org/10.1007/978-3-319-06028-6_8
7. Kato, M.P., Sakai, T., Yamamoto, T., Pavlu, V., Morita, H., Fujita, S.: Overview of the NTCIR-12 MobileClick-2 Task. In: Proceedings of the NTCIR-12, pp. 104–114 (2016)
8. Kelly, D., Arguello, J., Edwards, A., Wu, W.: Development and evaluation of search tasks for IIR experiments using a cognitive complexity framework. In: Proceedings of the ICTIR, pp. 101–110 (2015)
9. Li, J., Huffman, S., Tokuda, A.: Good abandonment in mobile and PC internet search. In: Proceedings of the the 32nd SIGIR, pp. 43–50 (2009)
10. Manabe, T., et al.: Information extraction based approach for the NTCIR-10 1CLICK-2 Task. In: Proceedings of the NTCIR-10, pp. 243–249 (2013)
11. Tojima, T., Yukawa, T.: Query classification system based on snippet summary similarities for NTCIR-10 1CLICK-2 task. In: Proceedings of the NTCIR-10, pp. 250–254 (2013)
12. Wildemuth, B., Freund, L., Toms, E.G.: Untangling search task complexity and difficulty in the context of interactive information retrieval studies. J. Doc. **70**(6), 1118–1140 (2014). https://doi.org/10.1108/JD-03-2014-0056
13. Yoshioka, M.: Query classification by using named entity recognition systems and clue keywords. In: Proceedings of the NTCIR-10, pp. 255–259 (2013)

A FAQ Search Training Method Based on Automatically Generated Questions

Takuya Makino[1,2(✉)], Tomoya Noro[1], Hiyori Yoshikawa[1,2], Tomoya Iwakura[1,2], Satoshi Sekine[2], and Kentaro Inui[2,3]

[1] Fujitsu Laboratories Ltd., Kawasaki, Japan
{makino.takuya,t.noro,y.hiyori,iwakura.tomoya}@jp.fujitsu.com
[2] RIKEN AIP, Tokyo, Japan
satoshi.sekine@riken.jp
[3] Tohoku University, Sendai, Japan
inui@ecei.tohoku.ac.jp

Abstract. We propose a FAQ search method with automatically generated questions by a question generator created from community Q&As. In our method, a search model is trained with automatically generated questions and their corresponding FAQs. We conducted experiments on a Japanese Q&A dataset created from a user support service on Twitter. The proposed method showed better Mean Reciprocal Rank and Recall@1 than a FAQ ranking model trained with the same community Q&As.

1 Introduction

Call centers usually prepare pairs of frequently asked questions and their answers as FAQs. FAQs are used by users as target documents from a search engine and by call centers as references to replay to customers' questions.

Each user usually searches FAQs with a query such as keywords or a natural sentence. One of the problems in FAQ searches is that vocabularies and writing style of users differ from those of FAQs. For example, if a user uses the query "Could not see HP" for a search, an answer FAQ "Internet connection problems for browsing home pages" would not be a result because the words in the query are not included in the FAQ. A possible solution for the problem is training a search model with queries given by users and their corresponding FAQs. However, there is usually no such data at early stages of an operation.

We propose a method for training a search model with automatically generated questions. Figure 1 shows an overview of our method. Our method trains a question generator with online Q&As. Then, various questions for the answer of each FAQ are generated by the question generator. Finally, our method trains a search model with the automatically generated questions of FAQs as queries. By using a question generator trained with online Q&As in which vocabularies and writing style are similar with users of a target FAQ search, we can expect to obtain questions that consist of the users' vocabularies.

© Springer Nature Switzerland AG 2018
Y.-H. Tseng et al. (Eds.): AIRS 2018, LNCS 11292, pp. 67–73, 2018.
https://doi.org/10.1007/978-3-030-03520-4_7

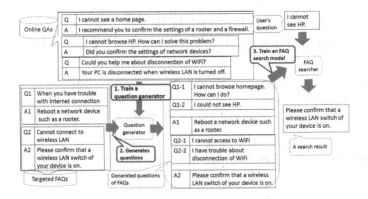

Fig. 1. An overview of the proposed method.

We evaluate our method on a data set created from a Japanese user support service for PC products on Twitter. The experimental results show that our method trained with questions generated by the question generator shows better MAP than a search model trained with questions and answers from Yahoo! Chiebukuro[1].

2 Related Work

Makino et al. [5] proposed use of query logs in a corresponding history for training by assigning the query logs to FAQs. The proposed method is similar in use of automatically generated training data. However, the proposed method generates training data from corresponding histories of a target domain. In contrast, our method generates new queries and can be used without corresponding histories.

Use of machine translation for acquiring different expressions was also proposed. Dong et al. [2] proposed a machine translation based method for generating paraphrases of a given question. The method obtains different sentences by translating an original sentence into a foreign language and then translating the translated back one into the language of the original sentence. This method is complementary with our method.

Serban et al. [6] proposed generating questions for factoid question answering; however, our focus is non-factoid one. Duan et al. [3] proposed question generation methods for question answering. Their methods consists of a combination of question template generation and filling in the template. In contrast, our method directly generates questions for an answer.

Yahoo! Answers[2] was used in [7] for training an answer ranking engine. Compared with use of community Q&As such as Yahoo! Answers and Yahoo! Chiebukuro for training a search model, our method has the following advantages:

---

[1] https://chiebukuro.yahoo.co.jp/.
[2] https://answers.yahoo.com/.

1. Training with target domain FAQs. A search model can be trained on target FAQs with questions given by a question generator, which would consist of users' vocabularies.
2. Correctness of training data. Training data generated from community Q&As may contain appropriate answers in irrelevant documents. This is because community Q&As may contain Q&As about the same topic. In contrast, FAQs created by a call center are not usually redundant.

3   Proposed Method

3.1   Learning a Question Generator

The proposed method first trains a question generator with Q&As from a community Q&A site. We used a sequence-to-sequence model [8] for training a question generator. Each Q&A is a thread that consists of its question and its answers. We extract pairs of the sequence of words in a question $\mathbf{y} = \langle y_1, ..., y_M \rangle$ and in a best-answer $\mathbf{x} = \langle x_1, ..., x_N \rangle$ as training data from each thread for training a question generator. A best-answer was selected by the questioner who created a thread. Let $D = \{(\mathbf{x}_1, \mathbf{y}_1), ..., (\mathbf{x}_K, \mathbf{y}_K)\}$ be the training data. The learning problem is to generate $\mathbf{y}_i$ from $\mathbf{x}_i$ in each D. The objective function to be minimized is the following:

$$L(\phi) = \sum_{(\mathbf{x}, \mathbf{y}) \in D} - \log p(\mathbf{y}|\mathbf{x}; \phi).$$

The $p(\mathbf{y}|\mathbf{x}; \phi)$ is the probability of generating $\mathbf{y}$ from $\mathbf{x}$ defined as:

$$p(\mathbf{y}|\mathbf{x}; \phi) = \prod_{t=1}^{M} p(y_t|y_1, ..., y_{t-1}; \mathbf{x}; \phi).$$

3.2   Generating Training Data with a Question Generator

Let the i-th FAQ in a FAQ collection be d_i that include the question q_i and the answer a_i. For generating questions for d_i, the answer part a_i is fed into a question generator. A set of b questions $\{q_i^1, ..., q_i^b\}$ for a_i is generated with the question generator using beam search. Each question-answer pair in $\{q_i^1, ..., q_i^b\}$ and d_i is appended to training data as $R = \{(q_1^1, d_1), ..., (q_1^b, d_1), ..., (q_i^1, d_i), ..., (q_i^b, d_i), ...\}$ for training the FAQ search model.

3.3   Learning a FAQ Search Model

An FAQ search model returns the FAQ that has the highest score in a collection of FAQs S for a given query q:

$$d = \operatorname*{argmax}_{d' \in S} F_\theta(q, d'),$$

where $F_\theta(q, d')$ is the score of a FAQ d' for a question q calculated with learnable parameter θ.

In the training phase with a training set R, the objective is to obtain the parameter θ that minimizes the following:

$$\theta = \operatorname*{argmin}_{\theta'} E(\theta'), \quad E(\theta') = \sum_{(q^{(i)}, d^{(i)}) \in R} \left\{ \max(0, l_{\theta'}(q^{(i)}, d^{(i)})) \right\},$$

$$l_{\theta'}(q^{(i)}, d^{(i)}) = \max_{d' \in S(q^{(i)})} F_{\theta'}(q^{(i)}, d') - F_{\theta'}(q^{(i)}, d^{(i)}),$$

where $S(q^{(i)})$ is the negative sample FAQs for $q^{(i)}$. $E(\theta')$ is 0 if a relevant FAQ is ranked higher than all the other irrelevant FAQs for all the questions in the training data.

3.4   FAQ Search Model

We use Supervised Semantic Indexing (SSI) [1] for training FAQ search models. SSI calculates the score of a document $\mathbf{d}$ for a query $\mathbf{q}$ as follows:

$$F_{\mathbf{U}, \mathbf{V}}(\mathbf{q}, \mathbf{d}) = \mathbf{q}^\top (\mathbf{U}^\top \mathbf{V} + \mathbf{I}) \mathbf{d},$$

where $\mathbf{U}, \mathbf{V} \in \mathbb{R}^{K \times N}$ are learnable parameters and $\mathbf{I}$ is an identical matrix. $\mathbf{q}, \mathbf{d} \in \mathbb{R}^N$ are vectors that denote occurrences of words in a query and a document respectively. q_i and d_i correspond to the i-th word of $\mathbf{q}$ and the i-th word of $\mathbf{d}$. A query and a document are mapped to K-dimensional vectors obtained by summing the corresponding vectors of words of $\mathbf{U}$ for the query and $\mathbf{V}$ for the document respectively. The occurrence frequency of the i-th word is a normalized sum of TFIDF scores of all the words:

$$q_i = \frac{tfidf(i, \mathbf{q}, C_q)}{\sum_{j=1}^{|\mathbf{q}|} tfidf(j, \mathbf{q}, C_q)},$$

where $tfidf(i, \mathbf{q}, C_q)$ is the tfidf score of the i-th word in $\mathbf{q}$. The document frequency of the i-th word in $\mathbf{q}$ is calculated based on a collection of question parts of FAQs C_q. d_i is calculated in the same manner as q_i with a collection of answer parts of FAQs C_a.

We set the dimensions of word vectors in $\mathbf{U}$ and $\mathbf{V}$ to 100. We choose the top-10,000 words for learning word vectors based on their frequencies in the training data.

4   Experimental Settings

4.1   Dataset

We created an evaluation dataset of replies from the @Fujitsu_FMV_QA account to users. Figure 2 shows an image of a reply from the @Fujitsu_FMV_QA account to a Twitter user. @Fujitsu_FMV_QA replies to Twitter users who are having trouble with their PC such as Internet disconnection. Replies from @Fujitsu_FMV_QA sometimes refer to FAQs about PC products. We extract a tweet from a user as a question and the FAQ that is referred to in the @Fujitsu_FMV_QA's reply to the tweet as an answer.

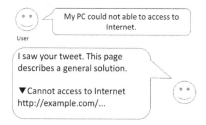

Fig. 2. An image of a reply of @Fujitsu_FMV_QA to a Twitter user

We compared search models trained with the following datasets:

FAQ-only: Training data generated from FAQs in the target task. The question of a FAQ is used as a query, and the answer part of the FAQ is used as its relevant document. Answer parts of all other FAQs are used as irrelevant documents.

Y!: Training data generated from Yahoo! Chiebukuro. We extract pairs of a question and its best-answer from the following categories: "Software", "Internet" and "PC". A question is used as a query and its best-answer is used as the relevant document. Randomly sampled five best-answers are used as irrelevant documents for each query.

QGen: Training data generated by the proposed method. We trained a question generator with Y!. We used a recurrent neural network-based encoder-decoder with an attention mechanism [4]. For generating questions, we used beam search. The beam width b is set to 10 and 50 respectively. Each generated question is used as a query for a search, and the question of a corresponding FAQ is used as a relevant document. Questions of 5 randomly sampled FAQs other than a relevant FAQ are used as irrelevant documents for each query.

TW: Training data generated from reply pairs of Twitter. A tweet of a user is used as a query and the FAQ in the reply to the query by @Fujitsu_FMV_QA is used as a relevant document. Questions from all other FAQs are used as irrelevant documents.

The 1,133 FAQs referred to in replies by @Fujitsu_FMV_QA in the Twitter data were selected as target documents for a search. To segment words from Japanese text, we used MeCab [3]. We evaluate search models with Mean Reciprocal Rank (MRR), Recall@1 and Recall@10. Table 1 shows the size of each dataset.

---

[3] http://taku910.github.io/mecab/.

Table 1. The size of each dataset.

Dataset	$|R|$
FAQ	12,655
Y!	393,873
QGen (b=10)	131,280
QGen (b=50)	656,400
Tw	2,639
Test data	565
Development data	565

5   Experimental Results

Table 2 shows the results of each method. The proposed method that uses automatically generated questions shows better MRR and R@1 tha FAQ-only and Y!. In addition, by increasing the number of generated queries, accuracy was improved. We think one reason is variation of generated questions. Generated questions for a FAQ have almost the same words but are slightly different from each other. This is because beam search-based generation tends to assign high scores to similar questions. However, with the increase of the number of generated questions, we think the accuracy was improved by training with more questions. R@10s for QGen (b=10) and QGen (b=50) are lower than Y!. We think Y! include more questions than our method and therefore leads to a higher R@10. However, our method' MRR score is still higher than tfidf, FAQ-only and Y!.

Table 2. Experimental results.

Method	Training data	MRR	R@1	R@10
tfidf	-	0.042	0.009	0.099
SSI	FAQ	0.044	0.012	0.101
SSI	Y!	0.096	0.060	0.154
SSI	QGen (b=10)	0.098	0.065	0.152
SSI	Y!+QGen (b=10)	0.100	0.067	0.154
SSI	QGen (b=50)	0.081	0.055	0.127
SSI	Y!+QGen (b=50)	0.097	0.067	0.138
SSI	TW	0.291	0.216	0.423

SSI with TW shows the best accuracy. One of the reasons is the following: a subset of FAQs is frequently referred to by the support account on Twitter and most of FAQs are included in TW. Another reason is vocabularies of TW are similar with those of the test data. However, the result can only be obtained

Table 3. Examples of top ranked FAQs by search models for query "I have trouble. I cannot hear any sounds from my PC."

SSI trained with Y!	[Update Navi] Please tell me how to solve a problem. I'm in trouble
SSI trained with QGen (b=50)	I cannot hear any sounds from a speaker

when we have questions from users and their corresponding answers. In contrast, our method can be applied without such data. Therefore, our method is useful for preparing an initial search model at an early stage.

Table 3 shows a successful example with the proposed method. Since the proposed method generated questions such as "No sounds when I turn on PC" for the FAQ, the relevant FAQ is ranked higher than the other FAQs for the question in the example.

6 Conclusion

We proposed a training method for a search with automatically generated questions. The experimental results on data of user support services showed that our method contributed to improved accuracy. We would like to develop a method that considers diversity when generating questions by a question generation model in future work.

References

1. Bai, B., et al.: Supervised semantic indexing. In: Proceedings of CIKM 2009 (2009)
2. Dong, L., Mallinson, J., Reddy, S., Lapata, M.: Learning to paraphrase for question answering. In: Proceedings of EMNLP 2017, pp. 875–886 (2017)
3. Duan, N., Tang, D., Chen, P., Zhou, M.: Question generation for question answering. In: Proceedings of EMNLP 2017, pp. 866–874. Association for Computational Linguistics (2017)
4. Luong, T., Pham, H., Manning, C.D.: Effective approaches to attention-based neural machine translation. In: Proceedings of EMNLP 2015, pp. 1412–1421. Association for Computational Linguistics (2015)
5. Makino, T., Noro, T., Iwakura, T.: An FAQ search method using a document classifier trained with automatically generated training data. In: Booth, R., Zhang, M.-L. (eds.) PRICAI 2016. LNCS (LNAI), vol. 9810, pp. 295–305. Springer, Cham (2016). https://doi.org/10.1007/978-3-319-42911-3_25
6. Serban, I.V., et al.: Generating factoid questions with recurrent neural networks: The 30m factoid question-answer corpus. In: Proceedings of ACL 2016, pp. 588–598. Association for Computational Linguistics (2016)
7. Surdeanu, M., Ciaramita, M., Zaragoza, H.: Learning to rank answers on large online QA collections. In: Proceedings of ACL 2008, pp. 719–727 (2008)
8. Sutskever, I., Vinyals, O., Le, Q.V.: Sequence to sequence learning with neural networks. In: Proceedings of NIPS 2014, pp. 3104–3112 (2014)

Embedding

A Neural Labeled Network Embedding Approach to Product Adopter Prediction

Qi Gu[1,2], Ting Bai[1,2], Wayne Xin Zhao[1,2(✉)], and Ji-Rong Wen[1,2]

[1] School of Information, Renmin University of China, Beijing, China
{guqi,baiting}@ruc.edu.cn, batmanfly@gmail.com, jirong.wen@gmail.com
[2] Beijing Key Laboratory of Big Data Management and Analysis Methods, Beijing, China

Abstract. On e-commerce websites, it is common to see that a user purchases products for others. The person who actually uses the product is called *the adopter*. Product adopter information is important for learning user interests and understanding purchase behaviors. However, effective acquisition or prediction of product adopter information has not been well studied. Existing methods mainly rely on explicit extraction patterns, and can only identify exact occurrences of adopter mentions from review data. In this paper, we propose a novel Neural Labeled Network Embedding approach (NLNE) to inferring product adopter information from purchase records. Compared with previous studies, our method does not require any review text data, but try to learn effective prediction model using only purchase records, which are easier to obtain than review data. Specially, we first propose an Adopter-labeled User-Product Network Embedding (APUNE) method to learn effective representations for users, products and adopter labels. Then, we further propose a neural prediction approach for inferring product adopter information based on the learned embeddings using APUNE. Our NLNE approach not only retains the expressive capacity of labeled network embedding, but also is endowed with the predictive capacity of neural networks. Extensive experiments on two real-world datasets (*i.e.,* JingDong and Amazon) demonstrate the effectiveness of our model for the studied task.

Keywords: Labeled network embedding · Product adopters
Neural network · e-commerce

1 Introduction

With the increasing popularity of online e-commerce services, purchasing online has become a shopping tendency in daily life. An interesting phenomenon is that a user buys products not only for herself but also for others. For example, it's common to see that a mother buys products for her daughter or husband. Given a purchase record, the person who buys the product is called *the buyer*, while the person who actually uses the product is called *the adopter* [20]. Such a purchase behavior can be easily inspected from review data. Take an online review as an

© Springer Nature Switzerland AG 2018
Y.-H. Tseng et al. (Eds.): AIRS 2018, LNCS 11292, pp. 77–89, 2018.
https://doi.org/10.1007/978-3-030-03520-4_8

example: "I bought my daughter this book." This review has revealed that the product buyer and the adopter are not the same person. We can easily infer the actual product adopter of the book is the daughter of the buyer.

Compared with the buyer, the adopter is likely to have a different shopping preference. While, many intelligent e-commerce systems, *e.g.,* recommender systems, usually ignore to capture such a preference drift in purchase behaviors, which may result in unsuitable response. Previous studies have also shown that product adopter information is important in various e-commerce applications, such as user profiling [20, 24] and product recommendation [23].

As illustrated in the above example, for extracting product adopter information, existing methods mainly utilize the self-disclosure behavior of users in online reviews [20, 23]. However, it has been reported in [23] that only about 10% online reviews have contained an adopter-related mention. Furthermore, many users only rate products, but don't give any textual comments to products. Worst of all, some reviews are copied or scribbled. These difficulties make it usually infeasible or unreliable for large-scale adopter detection with review data. To address these difficulties, this work aims to automatically infer the adopter information using only purchase records without review text. Our idea is that a purchase behavior itself is a combination of the involved user (*i.e.,* buyer), adopter and product. A user is likely to show varying preferences when she buys products for different adopters (including herself), which further forms different user-adopter-product combinations. If we could mine and learn effective combination characteristics from purchase records, it would be able to infer the adopter information given a purchase behavior.

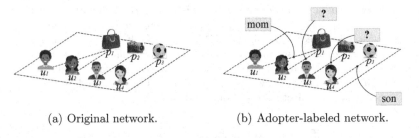

(a) Original network.   (b) Adopter-labeled network.

Fig. 1. An illustrative comparison of unlabeled and labeled user-product networks. The labeled $u_1 \rightarrow_{mom} p_1$ indicates a man u_1 buys a bag p_1 for his mother.

Previously, a purchase behavior is usually modeled as a two-way interaction between a user and a product, and we can use a bipartite network (see Fig. 1a) to characterize purchase records. To further incorporate adopter information, we propose to associate each edge with an adopter label. As shown in Fig. 1(b), we call such a network *Adopter-labeled User-Product Network (AUPN)*. Our task becomes how to infer the missing adopter labels for unlabeled edges. Although AUPN is more suitable to incorporate the adopter information, it increases the complexity to characterize the underlying network structure. Inspired by the recent progress on network embedding [6,14,18], we propose to learn vector-ized node representations (called *embeddings*) for capturing effective informa-tion characteristics of buyers, products and adopters. Embeddings can encode complex structure information and provide a compact information representa-tion method. Although network embedding receives much attention from the research community, adopter-labeled network embedding has been seldom stud-ied. Hence, existing methods are not able to model adopter-labeled information.

In this paper, we first propose a new Adopter-labeled User-Product Net-work Embedding (AUPNE) method to embed user vertices, product vertices and adopter labels into the same latent space. Our embedding method is able to bet-ter characterize the three-way user-adopter-product combinations. For inferring the product adopter, we further propose a Neural Labeled Network Embedding (NLNE) approach by utilizing the learned embeddings from AUPNE. Our NLNE approach not only retains the expressive capacity of labeled network embedding, but also is endowed with the predictive capacity of neural networks. Extensive experiments on two real-world datasets (*i.e.*, JingDong and Amazon) demon-strate the effectiveness of our model for the studied task.

2   Preliminaries

On an e-commerce website, let $\mathcal{U}$ denotes a user set and $\mathcal{P}$ denotes a product set. It is common for users to purchase products for others. The person who actually uses the product is called the adopter. We assume that an adopter label set $\mathcal{A}$ is given. Given a user $u \in \mathcal{U}$ and a product $p \in \mathcal{P}$, the adopter information of a purchase record $\langle u, p \rangle$ can be assigned with an adopter label $a \in \mathcal{A}$, where a indicates the adopter who actually uses the product.

With these notations, an *Adopter-labeled User-Product Network (AUPN)* can be formally defined as $\mathcal{G} = (\mathcal{V}, \mathcal{E}, \mathcal{A}, \mathcal{Y})$, where $\mathcal{V} = \mathcal{U} \cup \mathcal{P}$ is the union of user set $\mathcal{U}$ and product set $\mathcal{P}$, $\mathcal{E} \subseteq \mathcal{U} \times \mathcal{P}$ is the set of edges between user vertices and product vertices, and $\mathcal{Y} \subseteq \mathcal{U} \times \mathcal{P} \times \mathcal{A}$ is the set of adopter-labeled edges. An example of AUPN is illustrated in Fig. 1(b), where we have user set $\mathcal{U} = \{u_1, u_2, u_3, u_4\}$ and product set $\mathcal{P} = \{p_1, p_2, p_3\}$. We have two labeled edges with the adopter labels of "mom" and "son" respectively.

Based on the above definitions, the task of *product adopter prediction* aims to infer the adopter label $a \in \mathcal{A}$ given a purchase record $\langle u, p \rangle$.

In what follows, we will introduce the Neural Labeled Network Embedding (NLNE) approach to product adopter prediction in detail. We first introduce a new labeled network embedding method in Sect. 3, and then give a neural prediction method based on the learned embeddings in Sect. 4.

3   Adopter-Labeled User-Product Network Embedding

A purchase behavior can be considered as a combination among the user, adopter and product. We would like to learn effective information representations for the three involved elements from purchase records. The learned representations will be subsequently used by the proposed neural prediction method. Given an AUPN $\mathcal{G} = (\mathcal{U} \cup \mathcal{P}, \mathcal{E}, \mathcal{A}, \mathcal{Y})$, the aim of network embedding is to characterize user vertices, product vertices as well as adopter labels with d-dimensional vectors, namely $\{r_u\}_{u \in \mathcal{U}}$, $\{r_p\}_{p \in \mathcal{P}}$ and $\{r_a\}_{a \in \mathcal{A}}$. Here, $r_u \in \mathbb{R}^d$, $r_p \in \mathbb{R}^d$ and $r_a \in \mathbb{R}^d$ denote the embedding vectors of user u, product p and adopter a respectively. We call our proposed model *Adopter-labeled User-Product Network Embedding (AUPNE)*. Recall that in our setting each purchase record corresponds to an edge in the network, which is either unlabeled or adopter-labeled. Next, we present our method in the two cases.

3.1   Modeling Unlabeled Edges

We first study how to model unlabeled edges, where we do not know the actual adopter for a purchase record. Following the idea in LINE [18], we characterize the conditional probability of a product p given a user u, denoted by $P(p|u)$. To model $P(p|u)$, we adopt a softmax function to map the inner-product values between embedding vectors into probability distributions:

$$P(p|u) = \frac{\exp\left(r_p^\top \cdot r_u\right)}{\sum_{p' \in \mathcal{P}} \exp\left(r_{p'}^\top \cdot r_u\right)}. \tag{1}$$

Finally, we maximize the probability of all unlabeled purchase records to learn user representations $\{r_u\}_{u \in \mathcal{U}}$ and product representations $\{r_p\}_{p \in \mathcal{P}}$. Here, we equivalently minimize its negative logarithm:

$$O_{unlabeled} = - \sum_{\langle u,p \rangle \in \mathcal{E}} \log P(p|u). \tag{2}$$

3.2   Modeling Adopter-Labeled Edges

The above embedding method directly drives the representations of a user and her purchased products to be close. Indeed, most of the existing recommendation algorithms [8,16] also hold the similar assumption, in which adopter information is not considered. Since the buyer and the adopter may not be the same person, simply considering the user-product interactions in the unlabeled network may not be sufficient for modeling complicated purchase behaviors. It is intuitive that the user (*i.e.*, buyer) and adopter will jointly affect the final decision. We have to consider the preferences of the user herself and the adopter for better characterizing purchase behaviors.

Our idea is that when a user buys products for others, her shopping preference will be drifted to the preference of the adopter. The final preference will be a combination of the two kinds of preferences. In this way, the product representation should be similar to the drifted user preference representation. To model the drifted user preference, we adopt a simple yet intuitive way by adding the embedding vectors of the user and the adopter, *i.e.*, $r_u + r_a$. Formally, given a labeled purchase record $\langle u, p, a \rangle$, we incorporate the adopter information to derive a adopter-oriented probability distribution $P_{(a)}(p|u)$:

$$P_{(a)}(p|u) = \frac{\exp\left(r_p^\top \cdot (r_u + r_a)\right)}{\sum_{p' \in \mathcal{P}} \exp\left(r_{p'}^\top \cdot (r_u + r_a)\right)}, \tag{3}$$

where the subscript a indicates the adopter information for $\langle u, p \rangle$. Compared with Eqs. 1 and 3 has a more intuitive explanation: the purchase behavior is determined jointly by the buyer and the corresponding adopter. Even for the same user, she is likely to show varying preferences when considering different adopters. Our formulation gives the flexibility to learn better user and product representations by incorporating adopter information.

For a totally labeled network, the objective function can be defined as:

$$O_{labeled} = - \sum_{\langle u, p, a \rangle \in \mathcal{Y}} \log P_{(a)}(p|u). \tag{4}$$

To learn the embeddings of our AUPN, we collectively model both unlabeled and labeled edges, and minimize the following objective function:

$$O = O_{unlabeled} + O_{labeled}, \tag{5}$$

where $O_{unlabeled}$ and $O_{labeled}$ are defined in Eqs. 2 and 4. The objective function Eq. 5 can be optimized in different ways [17]. One solution is to train the model with $O_{unlabeled}$ and $O_{labeled}$ simultaneously. An alternative solution is to learn $O_{unlabeled}$ first, and then fine-tune the embeddings with $O_{labeled}$. In our experiment, we adopt the second optimization way with the negative sampling method proposed in [12].

4   A Neural Prediction Approach to Inferring Product Adopter Information

Above, we have studied how to learn effective representations for users, products and adopters. These representations highly summarize the underlying network structure of our constructed AUPN. As introduced in problem statement in Sect. 2, the task of inferring adopter information aims to infer the adopter label a given a purchase record $\langle u, p \rangle$. In the following, we further introduce a neural prediction approach to accomplishing the given task based on the learned node embeddings.

Specially, we consider the studied task as a multi-classification problem, where we aim to assign an adopter label to an unlabeled purchase record. We would like to combine the expressive capacity of the labeled network embedding (Sect. 3) and the predictive capacity of neural networks. We first present an overall model sketch in Fig. 2. Our model is designed as a two-way interaction architecture, where the learned user embedding and product embedding are taken as input of the neural network. Furthermore, we enhance the user representation by incorporating the adopter preference. Next, we present the model details.

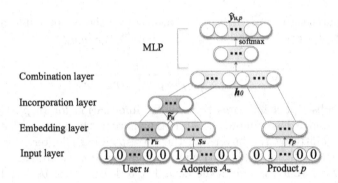

Fig. 2. A sketch of our proposed neural network model for product adopter prediction. Note that r_u, r_p, r_a are learned by our proposed embedding method AUPNE (Sect. 3).

Enhancing User Representations by Aggregating Associated Adopter Preferences. The preference of a user is likely to be affected by her associated adopters. Hence, for a user u, we first aggregate all historical adopter labels that have been used by u, denoted by $\mathcal{A}_u$. Furthermore, we average the embeddings of the adopter labels in $\mathcal{A}_u$ to derive a single representation s_u

$$s_u = \frac{1}{|\mathcal{A}_u|} \sum_{a \in \mathcal{A}_u} r_a, \tag{6}$$

where s_u characterizes the overall preference of all the associated adopters for u. The enhanced user representation $\widetilde{r}_u$ with adopter preference is defined as:

$$\widetilde{r}_u = r_u + s_u, \tag{7}$$

where such a representation is able to reflect the preference of the user herself and her associated adopters.

Product Adopter Prediction Through an MLP Classifier. Since a purchase record is associated with a user-product pair, we design a two-way neural network for product adopter prediction. Our task is considered as a multi-classification problem, in which each adopter is treated as a class label. The input vector h_0 of Multi-Layer Perceptron (MLP) [4] represents the edge $\langle u, p \rangle$ and it is composed by the user and product vectors:

$$h_0 = \text{combine}(\widetilde{r}_u, r_p), \tag{8}$$

where $\text{combine}(\cdot, \cdot)$ is an operation to combine the user and product vectors into a single vector. We consider three different combination methods, namely element-wise multiplication (M), addition (A) and concatenation (C). Then h_0 is fed into an MLP with L hidden layers. We can get the confidence vector for the classification task $\widehat{\boldsymbol{y}}_{u,p}$ by:

$$\widehat{\boldsymbol{y}}_{u,p} = \mathrm{softmax}(\boldsymbol{h}_L \cdot \boldsymbol{W}_L + \boldsymbol{b}_L)$$
$$\boldsymbol{h}_l = f(\boldsymbol{h}_{l-1} \cdot \boldsymbol{W}_{l-1} + \boldsymbol{b}_{l-1})$$
$$\cdots$$
$$\boldsymbol{h}_1 = f(\boldsymbol{h}_0 \cdot \boldsymbol{W}_0 + \boldsymbol{b}_0), \tag{9}$$

where $\boldsymbol{h}_l$ denotes the corresponding output of the l-th hidden layer, which is derived on top of the $(l-1)$-th hidden layer for $l \in \{1, ...L\}$, $f(\cdot)$ is an non-linear activation function implemented by the Rectifier Linear Unit (ReLU) [5], $\{\boldsymbol{W}_0, ..., \boldsymbol{W}_L\}$ are weight matrices, and $\{\boldsymbol{b}_0, ..., \boldsymbol{b}_L\}$ are bias terms.

The $|\mathcal{A}|$-dimensional probability distribution $\widehat{\boldsymbol{y}}_{u,p}$ is computed over the entire adopter set $\mathcal{A}$. The loss function is optimized by the cross entropy between the ground truth $\boldsymbol{y}_{u,p}$ and the predicted distribution $\widehat{\boldsymbol{y}}_{u,p}$:

$$L = - \sum_{\langle u,p,a \rangle \in \mathcal{Y}} \sum_{a' \in \mathcal{A}} y_{u,p,a'} \cdot \log(\widehat{y}_{u,p,a'}). \tag{10}$$

In this model, we utilize the expressive capacity of the adopter-labeled network embedding, since we take the learned representations as the input of our neural networks. Meanwhile, we also utilize the predictive capacity of deep learning by designing a two-way multi-layer neural networks. Specially, we also enhance the user representation by incorporate adopter preference.

5 Experiments

5.1 Experiment Setup

Datasets. We conduct experiments on two real-world datasets, *i.e.*, Jing-Dong [20] and Amazon [7], to verify the effectiveness of our proposed method for inferring adopter information. Each dataset contains a large number of purchase records, which consists of a user ID, product ID, and review text. We filter users and products with very few purchase records (frequency ≤ 20). In our task, we need to generate the ground-truth for adopter information of a purchase record. We first apply the bootstrapping algorithm [20] on review texts to learn adopter mentions and extraction patterns. Then, we manually check the validity of each adopter mention as an adopter label. We further manually merge adopter labels with same semantics, *e.g.*, *"gf"* with *"girl-friend"*. After that, we obtain 158 adopter labels from JingDong dataset, and 133 adopter labels from Amazon dataset (frequency ≥ 30). We use these adopter labels to generate labeled purchase records from review data by exact match. For each exact occurrence of an adopter label in a review, if the review matches one of the learned extraction patterns (*e.g.*, *"I buy $\cdots$ for $\cdots$"*), we form a user-product-adopter triplet, which is considered as a labeled edge. The statistics of our datasets are summarized in Table 1. We split the labeled edges with a ratio of 9:1 for training and test. All the unlabeled edges will not be split but used for training. Note all the review data is only used for generating ground-truth, and will be not used by our approach.

Table 1. Statistics of our datasets.

Datasets	$|\mathcal{U}|$	$|\mathcal{P}|$	$|\mathcal{E}|$	$|\mathcal{A}|$	$|\mathcal{Y}|$
JingDong	28,043	30,227	2,154,001	158	231,597
Amazon	25,831	35,915	294,940	133	99,173

Compared Algorithms. Our comparison methods include count-based methods (MostPopular, Product-based and User-based), unlabeled network embedding model (LINE) and our proposed approach.

- **MostPopular.** An adopter is ranked according to its total occurrence frequency.
- **Product-based.** An adopter is ranked by its occurrence frequency associated with product p.
- **User-based.** An adopter is ranked by its occurrence frequency associated with user u.
- **Product+User-based.** An adopter is ranked by a weighted liner combination of the frequencies obtained by *Product-based* and *User-based* methods.
- **LINE++.** We use LINE [18] with both first-order and second-order proximities to learn the user and product representations on our unlabeled networks. Then we obtain the representations of user and product by concatenating the corresponding vectors from the two modes. The derived representation is taken as input of an MLP component similar to our model.
- **AUPNE**$_{linear}$. After obtaining the embeddings with our AUPNE method, we adopt a simple linear ranking method, *i.e.*, $\text{score}(u, p, a) = \boldsymbol{r}_u^\top \cdot \boldsymbol{r}_a + \boldsymbol{r}_p^\top \cdot \boldsymbol{r}_a$.
- **NLNE.** It is our proposed neural network approach (Sect. 4) with the learned embeddings from AUPNE (Sect. 3). We set up three variants of NLNE, denoted by NLNE$_M$, NLNE$_A$ and NLNE$_C$, corresponding to element-wise multiplication, addition and concatenation respectively.

Evaluation Metrics. Given a purchase record, a comparison generates a list of adopter labels. We adopt two widely used ranking-based metrics, Mean Reciprocal Rank (MRR) and Hit ratio at rank k (Hits@k), to evaluate the performance of a ranked list.

Parameter Settings. We take 10% training data for parameter setting. For network embedding methods, the number of negative samples is set to 4. The number of hidden layers of the MLP is set to 2 and 1 for LINE++ and our AUPNE model, respectively. The dimension of embedding vectors is set to 512.

5.2 Results and Analysis

We present the results of MRR and Hit@k (*i.e.*, $k = 5$ and $k = 10$) on the product adopter prediction task in Table 2. We can have the following observations.

(1) For frequency-based baselines, MostPopular performs poorly by producing a global frequency-based ranked list. Product- and User-based methods overall

perform better than MostPopular by consider either user or product information. Surprisingly, the combination of Product- and User-based methods gives very competitive performance, even better than LINE++. An empirical explanation is that a user tends to repetitively incorporate same adopter labels, and a product tends to receive same adopter labels. Hence, frequency-based methods work very well when training data is sufficient.

(2) Our embedding method AUPNE with linear ranking function is better than LINE with an MLP-based ranking function. A major advantage of AUPNE over LINE++ is that it can incorporate adopter information and characterize three-way interaction, while LINE++ can only model two-way user-product interaction.

(3) Our NLNE methods perform very well. Especially, the variant with the concatenation mode NLNE$_C$ gives the best performance. Compared with the our linear method, NLNE further incorporates adopter preference and adopts a multi-layer neural network, which significantly improves the performance.

Table 2. Performance comparisons of the proposed methods and baselines.

Method	JingDong			Amazon		
	MRR	Hits@5	Hits@10	MRR	Hits@5	Hits@10
MostPopular	0.136	0.178	0.350	0.101	0.120	0.242
Product-based	0.225	0.350	0.441	0.081	0.112	0.120
User-based	0.375	0.628	0.639	0.245	0.356	0.366
Product+User-based	0.397	0.672	0.741	0.262	0.395	0.421
LINE++	0.469	0.623	0.741	0.180	0.305	0.459
AUPNE$_{linear}$	0.476	0.679	0.781	0.275	0.431	0.544
NLNE$_M$	0.474	0.636	0.758	0.168	0.207	0.340
NLNE$_A$	0.522	0.692	0.774	0.296	0.446	0.558
NLNE$_C$	**0.523**	**0.699**	**0.794**	**0.314**	**0.471**	**0.592**

To give a better understanding of why our method works well, we present a qualitative analysis of the learned embeddings of adopter labels for our AUPNE model on Amazon dataset. We first select five target adopter labels, and then recall the five most similar adopter labels using cosine distance. In Table 3, we can see the top ranked adopter labels are indeed very related to the target one.

5.3   Parameter Tuning

In this part, we construct the detailed analysis of the proposed model, including the number of hidden layers and embedding dimensions, and the percentage of labeled edges. At each time, we fix the other factors as the optimal settings and tune the studied factor. To verify the effectiveness, we select two best baselines for comparison, namely Product+User-based method and LINE++. We report

Table 3. Examples of top five most similar adopter labels identified by our embedding method on the Amazon dataset.

Target adopter	Top five most similar adopter labels
mother	mother-in-law, grandmother, godmother, lady, stepmother
father	father-in-law, grandfather, stepfather, professor, parent
stepson	son-in-law, daughter-in-law, stepdaughter, grandbaby, godson
infant	newborn, great-granddaughter, great-grandson, godson, toddler
office	manager, staff, boss, dorm, partner

the results on JingDong dataset, while the results on Amazon dataset are similar and omitted here.

Effects of Deep Architecture. In Fig. 3(a), we can see that our NLNE outperforms all baselines with different numbers of hidden layers. This indicates that our AUPNE method is powerful to model the adopter information. Overall, our NLNE method reaches a relatively high performance with one or more hidden layers, which may indicate that our current task does not require very complicated neural network.

Effects of Embedding Dimension. We vary the number of embedding dimensions in $\{32, 64, 128, 256, 512, 1024\}$ for both LINE++ and our method. In Fig. 3(b), we can see that our NLNE method outperforms the baseline methods with different numbers of embedding dimensions.

Effects of Labeled Data. To examine the effect of the percentage of labeled edges, we first fix the test set, and then gradually increase the percentage of labeled edges in training data from 20% to 100% with a gap of 20%. In Fig. 3(c), we can see that the performance of all methods improves as the amount of labeled data increases. Our NLNE method consistently outperforms the other methods with all different percentages of labeled data.

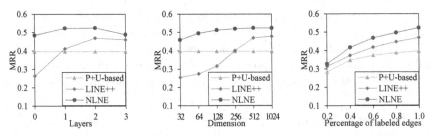

(a) Varying the number of layers. (b) Varying the number of dimensions. (c) Varying the percentage of labeled edges.

Fig. 3. Performance tuning with the varying of hidden layers, embedding dimensions, and percentage of labeled edges on the JingDong dataset (measured by MRR).

6   Related Work

Our work is closely related to the following studies in two aspects.

Mining E-Commerce Data. Our work is closely to e-commerce data mining and related applications. One of the most typical applications is product recommender systems [1, 2, 9], which learns user preference from historical purchase records. Recently, online review data mining has become a focus of both research and industry communities, and many studies utilize review data for various applications, including sales prediction [10], product ranking [11], rating prediction [3] and sentiment analysis [13, 22]. The two most related studies to ours are the extraction and utilization of adopter information from online review data [20, 23]. However, both studies rely on explicit patterns for extracting adopter mentions from review data, and their focus is to incorporate adopter information into matrix factorization framework for recommendation. As a comparison, our work aims to infer product adopters from purchase records with a neural approach.

Network Embedding. Network embedding aims to map nodes in a network to low-dimensional representations and effectively preserve the network structure. Most of network embedding studies attempt to preserve the network structural information, *e.g.,* Deepwalk [14], LINE [18], and node2vec [6]. Besides network topology, some studies utilized rich side information of networks, such as node content or labels in information networks [17, 19], node and edge attributes in social networks [21]. For e-commerce data, previous study has enhanced the performance of candidate retrieval by employing network embeddings to produce relevant communities [15]. Our network embedding method is built on a new adopter-labeled user-product network, and our focus is product adopter prediction using network embedding, which is different from existing studies.

7   Conclusion

In this paper, we have presented a novel Neural Labeled Network Embedding (NLNE) approach to inferring product adopter information. To the best of our knowledge, this is the first time that the task of product adopter prediction has been solved by an effective neural approach. Our approach not only retains the expressive capacity of labeled network embedding, but also is endowed with the predictive capacity of neural networks. Extensive experiments on two real-world datasets (*i.e.,* JingDong and Amazon) have demonstrated the effectiveness of our model for inferring the purchase adopters.

Currently, review data is not used by our approach, but it will be useful to utilize such review text besides purchase records. As future work, we will study how to incorporate review text into the proposed neural networks. In addition, we will also investigate how to model the buyer's profile for more accurate prediction of product adopter information.

Acknowledgement. This work was partially supported by the National Natural Science Foundation of China under Grant No. 61502502, the National Basic Research 973 Program of China under Grant No. 2014CB340403 and the Beijing Natural Science Foundation under Grant No. 4162032. Ting Bai was supported by the Outstanding Innovative Talents Cultivation Funded Programs 2016 of Renmin University of China.

References

1. Bai, T., Wen, J.R., Zhang, J., Zhao, W.X.: A neural collaborative filtering model with interaction-based neighborhood. In: Proceedings of the 2017 ACM on Conference on Information and Knowledge Management, pp. 1979–1982. ACM (2017)
2. Chen, J., Zhang, H., He, X., Nie, L., Liu, W., Chua, T.S.: Attentive collaborative filtering: multimedia recommendation with item-and component-level attention. In: Proceedings of the 40th International ACM Special Interest Group on Information Retrieval Conference on Research and Development in Information Retrieval, pp. 335–344. ACM (2017)
3. Ganu, G., Elhadad, N., Marian, A.: Beyond the stars: improving rating predictions using review text content. In: International Workshop on the Web and Databases, vol. 9, pp. 1–6. Citeseer (2009)
4. Gardner, M.W., Dorling, S.: Artificial neural networks (the multilayer perceptron)-a review of applications in the atmospheric sciences. Atmos. Environ. **32**(14–15), 2627–2636 (1998)
5. Glorot, X., Bordes, A., Bengio, Y.: Deep sparse rectifier neural networks. In: Proceedings of the 14th International Conference on Artificial Intelligence and Statistics, pp. 315–323 (2011)
6. Grover, A., Leskovec, J.: node2vec: scalable feature learning for networks. In: Proceedings of the 22nd ACM Special Interest Group on Knowledge Discovery and Data Mining International Conference on Knowledge Discovery and Data Mining, pp. 855–864. ACM (2016)
7. He, R., McAuley, J.: Ups and downs: modeling the visual evolution of fashion trends with one-class collaborative filtering. In: Proceedings of the 25th International Conference on World Wide Web, pp. 507–517 (2016)
8. Koren, Y., Bell, R., Volinsky, C.: Matrix factorization techniques for recommender systems. Computer **8**, 30–37 (2009)
9. Li, J., Ren, P., Chen, Z., Ren, Z., Lian, T., Ma, J.: Neural attentive session-based recommendation. In: Proceedings of the 2017 ACM on Conference on Information and Knowledge Management, pp. 1419–1428. ACM (2017)
10. Liu, Y., Huang, X., An, A., Yu, X.: ARSA: a sentiment-aware model for predicting sales performance using blogs. In: Proceedings of the 30th Annual International ACM Special Interest Group on Information Retrieval Conference on Research and Development in Information Retrieval, pp. 607–614. ACM (2007)
11. McGlohon, M., Glance, N.S., Reiter, Z.: Star quality: aggregating reviews to rank products and merchants. In: International Association for the Advance of Artificial Intelligence Conference on Web and Social Media (2010)
12. Mikolov, T., Sutskever, I., Chen, K., Corrado, G.S., Dean, J.: Distributed representations of words and phrases and their compositionality. In: Advances in Neural Information Processing Systems, pp. 3111–3119 (2013)
13. Pang, B., Lee, L., et al.: Opinion mining and sentiment analysis. Found. Trends ® Inf. Retr. **2**(1–2), 1–135 (2008)

14. Perozzi, B., Al-Rfou, R., Skiena, S.: Deepwalk: online learning of social representations. In: Proceedings of the 20th ACM Special Interest Group on Knowledge Discovery and Data Mining International Conference on Knowledge Discovery and Data Mining, pp. 701–710. ACM (2014)
15. Ren, Z., He, X., Yin, D., de Rijke, M.: Information discovery in e-commerce (2018)
16. Rendle, S.: Factorization machines with libfm. ACM Trans. Intell. Syst. Technol. **3**(3), 57 (2012)
17. Tang, J., Qu, M., Mei, Q.: PTE: predictive text embedding through large-scale heterogeneous text networks. In: Proceedings of the 21th ACM Special Interest Group on Knowledge Discovery and Data Mining International Conference on Knowledge Discovery and Data Mining, pp. 1165–1174. ACM (2015)
18. Tang, J., Qu, M., Wang, M., Zhang, M., Yan, J., Mei, Q.: Line: large-scale information network embedding. In: Proceedings of the 24th International Conference on World Wide Web, pp. 1067–1077 (2015)
19. Tu, C., Zhang, W., Liu, Z., Sun, M., et al.: Max-margin deepwalk: discriminative learning of network representation. In: International Joint Conference on Artificial Intelligence, pp. 3889–3895 (2016)
20. Wang, J., Zhao, W.X., He, Y., Li, X.: Leveraging product adopter information from online reviews for product recommendation. In: International Association for the Advance of Artificial Intelligence Conference on Web and Social Media, pp. 464–472 (2015)
21. Yang, C., Liu, Z., Zhao, D., Sun, M., Chang, E.Y.: Network representation learning with rich text information. In: International Joint Conference on Artificial Intelligence, pp. 2111–2117 (2015)
22. Zhang, Y., Lai, G., Zhang, M., Zhang, Y., Liu, Y., Ma, S.: Explicit factor models for explainable recommendation based on phrase-level sentiment analysis. In: Proceedings of the 37th International ACM Special Interest Group on Information Retrieval Conference on Research and Development in Information Retrieval, pp. 83–92. ACM (2014)
23. Zhao, W.X., Wang, J., He, Y., Wen, J.R., Chang, E.Y., Li, X.: Mining product adopter information from online reviews for improving product recommendation. ACM Trans. Knowl. Discov. Data **10**(3), 29 (2016)
24. Zhao, X.W., Guo, Y., He, Y., Jiang, H., Wu, Y., Li, X.: We know what you want to buy: a demographic-based system for product recommendation on microblogs. In: Proceedings of the 20th ACM Special Interest Group on Knowledge Discovery and Data Mining International Conference on Knowledge Discovery and Data Mining, pp. 1935–1944. ACM (2014)

RI-Match: Integrating Both Representations and Interactions for Deep Semantic Matching

Lijuan Chen[1,2]([envelope]), Yanyan Lan[1,2], Liang Pang[1,2], Jiafeng Guo[1,2], Jun Xu[1,2], and Xueqi Cheng[1,2]

[1] CAS Key Lab of Network Data Science and Technology,
Institute of Computing Technology, Chinese Academy of Sciences, Beijing, China
chenlijuan@software.ict.ac.cn,
{lanyanyan,pangliang,guojiafeng,junxu,cxq}@ict.ac.cn
[2] University of Chinese Academy of Sciences, Beijing, China

Abstract. Existing deep matching methods can be mainly categorized into two kinds, i.e. representation focused methods and interaction focused methods. Representation focused methods usually focus on learning the representation of each sentence, while interaction focused methods typically aim to obtain the representations of different interaction signals. However, both sentence level representations and interaction signals are important for the complex semantic matching tasks. Therefore, in this paper, we propose a new deep learning architecture to combine the merits of both deep matching approaches. Firstly, two kinds of word level matching matrices are constructed based on word identities and word embeddings, to capture both exact and semantic matching signals. Secondly, a sentence level matching matrix is constructed, with each element stands for the interaction between two sentence representations at corresponding positions, generated by a bidirectional long short term memory (Bi-LSTM). In this way, sentence level representations are well captured in the matching process. The above matrices are then fed into a spatial recurrent neural network (RNN), to generate the high level interaction representations. Finally, the matching score is produced by a k-Max pooling and a multilayer perceptron (MLP). Experiments on paraphrasing identification shows that our model outperforms traditional state-of-the art baselines significantly.

Keywords: Deep semantic matching · Word level interactions Sentence level representations

1 Introduction

Matching two sentences is a core problem of many applications in natural language processing, such as information retrieval and question answering. Taking information retrieval as an example, given a query and a document, a matching

Y.-H. Tseng et al. (Eds.): AIRS 2018, LNCS 11292, pp. 90–102, 2018.
https://doi.org/10.1007/978-3-030-03520-4_9

function is created to determine the relevance degree between the query and the document.

Recently, deep neural networks have been applied in this area and achieved some progresses. These methods can be mainly categorized into two kinds: representation focused methods and interaction focused methods. Representation focused methods first encode each sentence as one dense vector, and then calculate the similarities of two sentences vectors as the matching score. Typical examples include ARC-I [3] and CNTN [8]. In general, this approach is straightforward and capable to capture the high level semantic meanings of each sentence. However, it will miss important detailed information by compressing such an entire sentence into a single vector. To tackle this problem, interaction focused methods turn to directly learn the interactions between two sentences. They first construct a word level matching matrix to capture detailed word level interaction signals. Then a deep neural network is applied on this matrix to abstract high level interaction signals. Finally, an MLP is used to calculate the matching score. State of the art methods include ARC-II [3], MatchPyramid [7], Match-SRNN [12], MV-LSTM [11], and BiMPM [14]. Interaction focused methods have the ability to integrate rich interaction signal, however, sentence level semantic meanings are not fully captured.

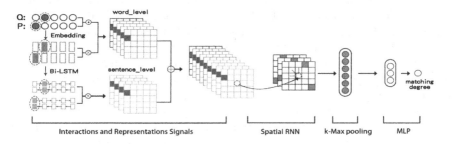

Fig. 1. An overview of RI-Match. ⊙(interaction in the word identities). ⊗(all interaction functions partly defined in word embedding level and sentence level).

Semantic matching is such a complex problem that both interaction signals and sentence representations need to be considered. In this paper, we propose a new deep architecture to integrate them, namely RI-Match[1]. For word level, we capture the interaction signals by the word identity and word embedding. For sentence level information, we adopt a Bi-LSTM to scan each sentence, and the sentence representations are obtained. Then a matrix can be constructed by computing the similarity between two sentence representations at corresponding positions. Finally, the above matrices are fed into a spatial RNN [12], which captures both nearly and long distant interactions. Furthermore, a k-Max pooling strategy [11] is adopted to select the top k strongest interaction signals, and a multi-layer perceptron(MLP) is utilized to obtain the matching score.

---

[1] We will release our code when our paper accepted.

We conduct two experiments on different tasks, such as paraphrasing identification and answer selection. The experimental results show that RI-Match outperforms traditional interaction and representation focused methods on paraphrasing identification tasks, demonstrating the advantage of combining the merits of both approaches.

2   Related Work

Existing deep learning methods for semantic matching can be mainly categorized into two kinds, i.e representation focused methods and interaction focused methods.

Representation focused methods represent two input sentences individually to a dense vectors in the same embedding space, and then define different functions to calculate the matching degree of the two sentence vectors. It is common to adopt the recursive neural network. The advantage of this method is to model complex and semantic phenomenon in the sentence level. It is easy to complete. However, the disadvantage is that the encoder would loss detail information of texts pair.

To tackle this problem, interaction focused methods has been proposed. It turns to capture more interactions relationship between two texts in word level. Then the matching degree can be determined by interaction matrix, which has achieved much attention, examples include MatchPyramid, MatchSRNN and MV-LSTM. Our method combines the merits of both deep matching methods.

The MatchPyramid constructs basic word level interaction matrix by defining different similarity functions. Then the model regards the semantic matching as image recognition by considering the matching matrix as one image. For this method, it will lose complex semantic matching information in the word level.

To overcome above defects, the same in the word level interaction, the Match-SRNN adopts neural tensor network to capture more complicated interactions [9]. Then the spatial RNN calculate the matching degree by the above interactions. This method partly solves the single matching matrix problem and output an interaction tensor. However, it cannot obtain the higher level interactions.

Different from the MatchPyramid and MatchSRNN, the MV-LSTM captures the interaction signals of texts in the sentence representations level. However, it is natural that the key words of texts sometimes partly determine the matching degree of sentences, this method cannot perfectly capture these signals.

For the above related work, it is natural for us to compose the semantic matching model with the different level signals, which takes both word level interaction signals and sentence level representations signals into account, namely RI-Match. It can richly capture the complex semantic matching information by feeding these two kinds level signals to next layer.

3   Method

In this section, we introduce our method which integrates both representation and interaction signals for deep semantic matching, namely RI-Match. As shown in Fig. 1, RI-Match consists of four components.

3.1   Interactions and Representations Structure

The goal of this component is to construct matching signals for two sentences in word interaction level and sentence representations level. Given two sentence $Q = (q_1, \cdots, q_m)$ and $P = (p_1, \cdots, p_n)$, where q_i and p_j denotes the i-th and j-th word in sentence Q and P. Sequence of word embeddings can be obtained by mapping each word identity into a vector, where we have $\mathbf{Q} = (\mathbf{q}_1, \cdots, \mathbf{q}_m)$ and $\mathbf{P} = (\mathbf{p}_1, \cdots, \mathbf{p}_n)$. In order to construct different level matching signals, we input both word identities Q and P, and word embeddings $\mathbf{Q}$ and $\mathbf{P}$.

Word Level Interactions Signal

The goal of this part is to represent word level interaction signals for sentences based on word identities and word embedding. We present several matching signals as a matching matrix $\mathbf{M}$, with each word-pair s_{ij} express the basic interaction between word q_i and p_j.

Based on Word Identities. It is natural for us to think that two sentences are more relevant if they contain more identical words.

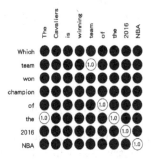

Fig. 2. Xnor operator, where the black circle elements are all value 0.

$Xnor$ can capture such information as follows,

$$s_0(q_i, p_j) = q_i \odot p_j. \tag{1}$$

where the $\odot$ stands for the $Xnor$ operation, which produce either 1 or 0 to measure whether two words are the same. We can visualize the matching matrix $\mathbf{M_0}$ in Fig. 2. As the example shows that both Q and P contain the words {$team$, 2016, NBA}, these partly determine the matching degree of two sentences.

However, the disadvantage of this similarity operator is that it cannot obtain the semantic matching situations, for example, the similar words matching signals are ignore. As we know, words 'won' and 'winning' have the similar meaning than words 'won' and '2016'. In order to capture word semantic similarities, we define similarity operators based on the word embeddings.

Based on Word Embedding. The word embedding is a fixed vector for every individual word, which is pre-trained by Glove. q_i and p_j stand for the i-th and j-th word representations in sentence Q and P.

Cosine is a common way to calculate similarity of two word embeddings, which regarded as the angle of two vectors. We show it as follows,

$$s_1(q_i, p_j) = \frac{q_i^T p_j}{\|q_i\| \cdot \|p_j\|}. \tag{2}$$

where $\| \cdot \|$ stands for the norm of the vector, we adopt L2 norm in this paper. Cosine function guarantees that exact matching signals will get the highest similarity scores 1.

Dot Product compares to cosine similarity operator, it takes the norm of word embeddings into account.

$$s_2(q_i, p_j) = q_i^T p_j. \tag{3}$$

The norm of word embedding can be interpreted as the importance of the word, for example, none word 'NBA' should be important than empty word 'the'.

Both cosine similarity and dot product similarity treat word similarity as a scale value s_1 and s_2 and obtain the matching matrix $\mathbf{M_1}$ and $\mathbf{M_2}$ shown in Fig. 3 while we also treat word similarity as a representation, such as a vector.

Element-Wise Multiplication is a direct way to combine the signals for two word vectors, and output a similarity representation of two words, which can be represented as follows,

$$s_3(q_i, p_j) = q_i \odot p_j. \tag{4}$$

where $\odot$ stands for element-wise multiplication. This method can produce the interaction matching vector matrix $\mathbf{M_3}$ which is different from the cosine and dot product, since they just obtain interaction matrix. This interaction signal can largely retain more details of sentences based on the word embedding.

Word Meaning Signal. Many deep models describe the word level interaction signals mostly in the interaction of two sentences. However, the meaning of word itself contains much useful information for semantic matching. For this purpose, we concatenate this signal with the above-mentioned word level interaction signals together to enrich the word level signals. First, we use the RNN to compress the dimension for every word embedding of each sentence, which can transform the word embedding to a dense vector in low dimension. Meanwhile, it can avoid a great difference between diverse signals in the dimension. It can

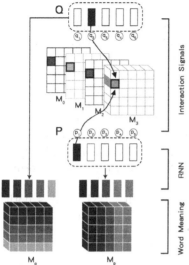

Fig. 3. Word level signals, where the matrix M_0, M_1, M_2, M_3 respectively are the result with operator of xnor, cosine, dot product and the element-wise multiplication.

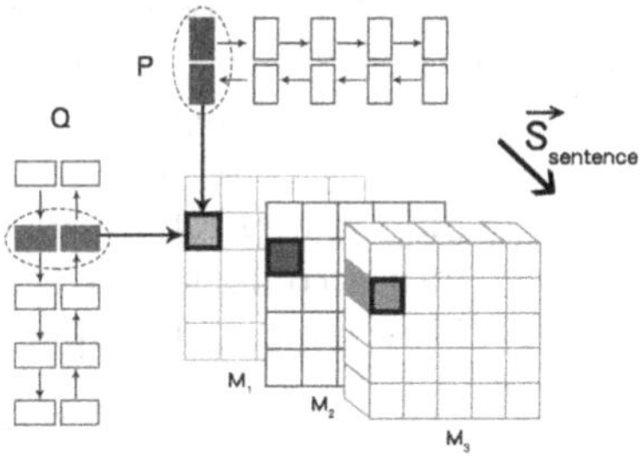

Fig. 4. Sentence level representations signals, where the matrix M_1, M_2, M_3 respectively are the result with operator of cosine, dot product and the element-wise multiplication.

also enrich the representations of every word. For the sentence $\mathbf{Q} = (\boldsymbol{q}_1, \ldots, \boldsymbol{q}_m)$ and $\mathbf{P} = (\boldsymbol{p}_1, \ldots, \boldsymbol{p}_n)$, it will get the new word representations as follows.

$$(\boldsymbol{q}_1, \ldots, \boldsymbol{q}_m) \xrightarrow{RNN} (\boldsymbol{q}_1^l, \ldots, \boldsymbol{q}_m^l),$$
$$(\boldsymbol{p}_1, \ldots, \boldsymbol{p}_n) \xrightarrow{RNN} (\boldsymbol{p}_1^l, \ldots, \boldsymbol{p}_n^l).$$

where the $\boldsymbol{q}_i^l$ and $\boldsymbol{p}_j^l$ separately stands for the new word representations for q_i and p_j, we show the RNN state in Fig. 3.

We cannot directly concatenate above representations with the word interaction signals. In the word interaction signals, it contains the $m \times n$ vectors, each vector represents every word-pairs interactions of two sentences. However for $\mathbf{Q}$ it contains m vectors and for $\mathbf{P}$ it contains n vectors. In order to concatenate word meaning signals with interaction signals, we stack every new word representations for n times to get the vector matrix M_q. For this method, We can get the vector matrix $\mathbf{M_p}$ in the similar way for m times as follows.

$$\boldsymbol{s}_q(ij), \forall j \in [1, n] \Leftrightarrow \boldsymbol{q}_i^l$$

$$\boldsymbol{s}_p(ij), \forall i \in [1, m] \Leftrightarrow \boldsymbol{p}_j^l$$

where the $\boldsymbol{s}_q(ij)$ is an vector in $\mathbf{M_q}$ shown in Fig. 3, $\boldsymbol{q}_i^l$ stacked with n times by row with the same color, therefore when $\forall j \in [1, n]$, they all stands for the $\boldsymbol{q}_i^l$. Similarly, we obtain $\mathbf{M_p}$ by stacking $\boldsymbol{p}_j^l$ with m times by column.

Finally, by concatenating the above-mentioned signals based on word identities and word embedding, we get the signals vector of sentences in word level as follows. We can see Fig. 3, it shows the detail of word level signals including word interactions signals and word meaning signals.

$$s_{word} = [s_0, s_1, s_2, s_3^T, s_q^T, s_p^T]^T \tag{5}$$

Sentence Level Representations Signal

The higher level matching signals are important to determine whether two sentences are match. The sentence representations signals depend on contextual information. Therefore, we adopt a parameter-shared bi-directional LSTM [4] to encode contextual embeddings to capture such information. It can well capture nearby words in the encode process.

For embedding matrix $\mathbf{Q} = (q_1, \ldots, q_m)$, bidirectional LSTM takes both previous and future context into account from two directions. Therefore, we utilize Bi-LSTM to process the input for each time-step as follows:

$$\begin{aligned}
\overrightarrow{h}_i^q &= \overrightarrow{LSTM}(h_{i-1}^q, q_i), i \in [1, m], \\
\overleftarrow{h}_i^q &= \overleftarrow{LSTM}(h_{i+1}^q, q_i), i \in [1, m].
\end{aligned} \tag{6}$$

Meanwhile, we apply the same BiLSTM to encode $\mathbf{P} = (p_1, \ldots, p_n)$:

$$\begin{aligned}
\overrightarrow{h}_j^p &= \overrightarrow{LSTM}(h_{j-1}^p, p_j), j \in [1, n], \\
\overrightarrow{h}_j^p &= \overleftarrow{LSTM}(h_{j+1}^p, p_j), j \in [1, n].
\end{aligned} \tag{7}$$

Therefore, we concatenate two vectors $\overrightarrow{h}_t$ and $\overleftarrow{h}_t$ together as $h = [\overrightarrow{h}_t^T, \overleftarrow{h}_t^T]^T$ for each position of sentence. It stands for t-th sentence representations from the two directions of the whole sentence. The $(\cdot)^T$ stands for the transposition operation. Then we can obtain the encode matrices as $\mathbf{Q_{Encode}} = (h_1^q, \ldots, h_m^q)$ and $\mathbf{P_{Encode}} = (h_1^p, \ldots, h_n^p)$ We adopt the interaction functions used in word level interaction signals, then we can obtain the sentence representations interaction vector as follows, the Fig. 4 shows the detail of sentence representations signals vector.

$$\begin{aligned}
s_1(h_i^q, h_j^p) &= \frac{h_i^{qT} h_j^p}{\|h_i^q\| \cdot \|h_j^p\|}, \\
s_2(q_i, p_j) &= h_i^{qT} h_j^p. \\
s_3(h_i^q, h_j^p) &= h_i^q \odot h_j^p, \\
s_{sentence} &= [s_1, s_2, s_3^T]^T.
\end{aligned} \tag{8}$$

Finally, we concatenate the word level interactions signals and sentence level representations signals together as follow.

$$s_{ij} = [s_{word}^T, s_{sentence}^T]^T. \tag{9}$$

where s_{ij} contains multiple signals in word interactions level and sentence representations level. For word level, it contains the word meaning signals and word interaction signals. Therefore, the output of this layer is tensor matrix of signals.

3.2 Spatial RNN

The second step is to apply spatial RNN to obtain the sentence interaction from multi-signals construction layer, from which we get multiple signals of two sentences in the word level and sentence representations level. Spatial RNN is a variation on the multi-dimensional RNN [1]. For the spatial RNN, given the interactions representations of prefixes $Q[1:i-1] \sim P[1:j], Q[1:i] \sim P[1:j-1]$ and $Q[1:i-1] \sim P[1:j-1]$, expressed as $\boldsymbol{h}_{i-1,j}, \boldsymbol{h}_{i,j-1}$ and $\boldsymbol{h}_{i-1,j-1}$, the interaction of prefixes $Q[1:i] \sim P[1:j]$ can be calculated by following equation:

$$\boldsymbol{h}_{ij} = f(\boldsymbol{h}_{i-1,j}, \boldsymbol{h}_{i,j-1}, \boldsymbol{h}_{i-1,j-1}, \boldsymbol{s}_{ij}). \tag{10}$$

where $\boldsymbol{s}_{ij}$ stands for the signals information from the multi-signals construction layer including word level interactions signals and sentence level representations signals.

We have many choices for function f. We adopt GRU since it has shown excellent performance in many tasks. In this paper, we use spatial RNN changed from traditional GRU. We extend it to spatial RNN as follows.

$$
\begin{aligned}
&\boldsymbol{q} = [\boldsymbol{h}_{i-1,j}^{\mathrm{T}}, \boldsymbol{h}_{i,j-1}^{\mathrm{T}}, \boldsymbol{h}_{i-1,j-1}^{\mathrm{T}}, \boldsymbol{s}_{ij}^{\mathrm{T}}]^{\mathrm{T}}, \\
&\boldsymbol{r}_l = \sigma(W^{(rl)}\boldsymbol{q} + \boldsymbol{b}^{(rl)}), \boldsymbol{r}_t = \sigma(W^{(rt)}\boldsymbol{q} + \boldsymbol{b}^{(rt)}), \\
&\boldsymbol{r}_d = \sigma(W^{(rd)}\boldsymbol{q} + \boldsymbol{b}^{(rd)}), \boldsymbol{r}^{\mathrm{T}} = [\boldsymbol{r}_l^{\mathrm{T}}, \boldsymbol{r}_t^{\mathrm{T}}, \boldsymbol{r}_d^{\mathrm{T}}], \\
&\boldsymbol{z}_i' = (W^{(zi)}\boldsymbol{q} + \boldsymbol{b}^{(zi)}), \boldsymbol{z}_l' = (W^{(zl)}\boldsymbol{q} + \boldsymbol{b}^{(zl)}), \\
&\boldsymbol{z}_t' = (W^{(zt)}\boldsymbol{q} + \boldsymbol{b}^{(zt)}), \boldsymbol{z}_d' = (W^{(zd)}\boldsymbol{q} + \boldsymbol{b}^{(zd)}), \\
&[\boldsymbol{z}_i, \boldsymbol{z}_l, \boldsymbol{z}_t, \boldsymbol{z}_z] = \text{SoftmaxByRow}([\boldsymbol{z}_i', \boldsymbol{z}_l', \boldsymbol{z}_t', \boldsymbol{z}_z']),
\end{aligned}
\tag{11}
$$

$$
\begin{aligned}
&\boldsymbol{h}_{ij}' = \phi(W\boldsymbol{s}_{ij} + U(\boldsymbol{r} \odot [\boldsymbol{h}_{i-1,j}^{\mathrm{T}}, \boldsymbol{h}_{i,j-1}^{\mathrm{T}}, \boldsymbol{h}_{i-1,j-1}^{\mathrm{T}}]^{\mathrm{T}}) + \boldsymbol{b}), \\
&\boldsymbol{h}_{ij} = \boldsymbol{z}_l \odot \boldsymbol{h}_{i,j-1} + \boldsymbol{z}_t \odot \boldsymbol{h}_{i-1,j} + \boldsymbol{z}_d \odot \boldsymbol{h}_{i-1,j-1} + \boldsymbol{z}_i \odot \boldsymbol{h}_{i,j}'.
\end{aligned}
\tag{12}
$$

where $U, W's$ and $b's$ are parameters, and SoftmaxByRow is a function to calculate softmax for every dimension by the four gates, as following:

$$[\boldsymbol{z}_p']_j = \frac{e^{[\boldsymbol{z}_p']_j}}{e^{[\boldsymbol{z}_p']_j} + e^{[\boldsymbol{z}_p']_l} + e^{[\boldsymbol{z}_p']_t} + e^{[\boldsymbol{z}_p']_d}}, p = i, l, t, d.$$

3.3 k-Max Pooling

Since spatial RNN getting the global interaction vectors between two texts, we introduce the third step to process such information by k-Max pooling.

The strong signals largely determine the matching degree of two sentences, these method has been approved to be valid in MV-LSTM. Therefore, we use k-Max pooling to automatically select top k strongest signals in the global interaction tensor, similar to [5]. Specifically for the spatial RNN matrix, we scan the matrix and directly return the top k values of every slice by the descending order to form a vector q.

3.4   MultiLayer Perception

Finally, we use a MLP to obtain the matching degree by aggregating the strong interaction information chosen by k-Max pooling, such information can be represented as vector q. For obtaining higher level representation r, vector q is feed into a full connection hidden layer. The final matching score can be obtained with a linear function:

$$r = f(W_s q + b_r), \quad s = W_s r + b_s.$$

where W_r and W_s donate the parameter metrics, and b_r and b_s are corresponding biases.

4   Experiments

In this section, we verify our model performance on two tasks: paraphrasing identification (PI) and answer sentence selection (ASS). We compare our model with state-of-the-art models on some standard benchmark datasets including Quora-question-pairs and WikiQA, to demonstrate the superiority of RI-Match against baselines.

4.1   Experimental Settings

First, we introduce our experimental settings, including parameter setting, and evaluation metrics.

Parameter Settings. We initialize word embeddings in the word embedding layer with 300-dimensional Glove word vectors pre-trained in the 840B Common Crawl corpus. On the paraphrasing identification task, we set the hidden dimension as 50 for Bi-LSTM and 40 for Spatial RNN. For the word meaning level, we set hidden size as 5 for the RNN, and set the top k as 5 for k-Max pooling, the learning rate is set to 0.001. On the answer selection task, we set the hidden dimension as 50 for Bi-LSTM and 10 for Spatial RNN. For the word meaning level, we set hidden size as 1 for the RNN, and set the topk as 10 for k-Max pooling, the learning rate is set to 0.0001. To train the model, we leverage Adam as our optimizer to update the parameters, and minimize the categorical cross entropy of the training set until the model convergence.

Evaluation Metrics. The PI is the binary classification problem, we adopt accuracy to evaluation the performance on this task. ASS can be considered as a ranking problem, we utilize mean average precision (MAP) and mean reciprocal rank (MRR).

where N is the number of testing ranking lists, M is the number of positive sentence in a ranking list. $S_j^{+(i)}$ is the j-th positive sentence in the i-th ranking list, $r(\cdot)$ denotes the rank of a sentence in the ranking list.

Table 1. Performance on Quora question dataset.

Models	Accuracy (%)
Siamese CNN	79.60
Multi-perspective-CNN	81.38
Siamese-LSTM	82.58
Multi-perspective-LSTM	83.21
L.D.C.	85.55
BiMPM-w/o-tricks	85.88
BiMPM-Full	**88.17**
RI-Match-WL	83.86
RI-Match-WL-w/o-cos	83.42
RI-Match-PSL	82.76
RI-Match-Full	**85.91**

Table 2. Performance on the WikiQA dataset.

Models	MAP	MRR
Word Count	0.652	0.665
ABCNN	0.692	0.711
Attention-CNN	0.689	0.696
Attention-LSTM	0.688	0.707
L.D.C.	0.705	0.723
GRU	0.659	0.669
BiMPM	**0.718**	**0.731**
RI-Match-Full	0.689	0.692

4.2 Paraphrasing Identification

Paraphrasing identification aims to determine whether two sentences tell the same story. In this Sub-section, we compare our model with relatively new baselines on the paraphrasing identification task.

Dataset. To evaluate the effectiveness of our model, we perform our experiments on the dataset of "Quora Question Pairs". To be a fair comparison, we adopt the splitting ways of [15]. This dataset consists of over 400,000 question pairs, and each question pair is annotated with a binary value indicating whether the two questions pairs are paraphrase of each other. The authors randomly select 5,000 paraphrases and 5,000 non-paraphrases as the dev set, and sample another 5,000 paraphrases and 5,000 non-paraphrases as the test set. Then they keep the remaining pairs as the training set. For getting the detail of dataset, please refer to [15][2].

Baseline. To make a sufficient comparison, we choose six relatively new baselines: Siamese CNN [6], Multi-Perspective CNN [2], Siamese-LSTM [10], Multi-Perspective-LSTM [2], L.D.C [16], BiMPM. For these baselines, they all adopt the above-mentioned splitting ways for the datasets.

- For the representation based methods, the Siamese CNN and Siamese-LSTM encode the sentence into dense vectors separately by CNN and LSTM. With the multi-perspective technique, the Siamese models are promoted as Multi-Perspective-CNN and Multi-Perspective-LSTM.
- For the interaction based methods, the L.D.C takes both the similarities and dissimilarities into account by decomposing and composing lexical semantics over texts. And the BiMPM guides the interaction with the attention-base neural architecture by encodes each sentence both in word embedding and

---

[2] We can obtain the source codes and dataset partition at: https://zhiguowang.github.io.

character embedding level. Both models have obtained the state-of-the art baselines in this datasets.

Performance Comparison. For a fair comparison with above baselines, we directly generate the result under the same setting from the literature. To be fair, we implement the BiMPM without the ticks (*e.g. character embedding, dropout, etc.*), we also list the baseline of BiMPM-Full with these tricks in Table 1. In order to show the influence of different level signals for semantic matching. We test the performance of RI-Match for it contains the signals just in the word level (RI-Match-WL) or the sentence representations level (RI-Match-PSL) and the full model (RI-Match-Full). The results are listed in Table 1. From the results, we could conclude the experimental findings as follows.

1. The models belong to interaction focused methods outperform the representation focused methods. This mainly emphasizes the importance of interaction structure. Representing the sentence to a vector directly will lost many information, this is an important factor that more deep models focus on describing the interaction of the texts.
2. Our model are better than all baselines, achieving the state-of-the-art performance. This illustrates the effectiveness of our model.
3. The RI-Match-Full is better than the RI-Match-WL and RI-Match-PSL, which shows that the signals both in word level and sentence representations level are important for semantic matching. We define more ways to construct the signals in word level rather than the sentence representations level, including word meaning signals and word level interaction signals. Therefore, we can see that RI-Match-WL is 1.1 point percentage higher than the RI-Match-PSL. This gap is obvious in this dataset. For further explanation for this, we check the performance just in word level without the cosine signals (RI-Match-WL-without-cos). We can see that RI-Match-WL is better than RI-Match-WL-without-cos. This means that we can improve performance of model in some extent by defining more different signals in each level.

4.3   Answer Selection

Answer selection is a task to rank the candidate answers based on their matching degree to the question. Evaluation metrics of this tasks are mean average precision (MAP) and mean reciprocal rank (MRR). We experiment on the WikiQA dataset.

Dataset. WikiQA is a public benchmark datasets, we need to rank the candidate answers according to a question. It includes 20,360 question-answer pairs in training set, 1,126 pairs in development set and 2,341 pairs in test set. We filter the questions without the correct answers.

Baseline. To make a fair comparison, we select following baselines.

- Word Count: is non-neural architecture, it calculates the frequency of non-stop words between question and answer.

- GRU: is used to obtain the sentence representations signals, it calculate the similarity of the sentence vector.
- ABCNN, BiMPM: ABCNN and BiMPM belong to the interaction focused methods introduced in related work, they are two state-of-art baselines for this task.
- Attention-based models: both the Attention-based-CNN [17] and Attention-based-LSTM [13] build the attention matrix after sentence representation, they adopt CNN and LSTM separately to encode sentences.

Performance Comparison. For a effective comparison, we report the results under the same setting from the literature. In the Table 2, we can see that RI-Match can do well on this task. Compared with the non-neural architecture, our models can automatically extract more semantic signals from the data, then leading the better performance. For BiMPM, they obtained their best performance by using the character embedding, it can richly capture the information in the input layer. The GRU belongs to the representation focused methods which encode the sentence by GRU. Both the L.D.C and BiMPM are belongs to interaction focused methods, they describe the interaction on the sentence representations level. From above, we can see that deep models basically beat the traditional method. In deep models, describing the interaction of texts sometimes can achieve better result. Our model can also do well in this task except the paraphrasing identification task.

5 Conclusions

In this paper, we propose a deep model by integrating both interactions and representations for deep semantic matching, namely RI-Match. We define various measure functions in each level to produce signals. Moreover, we adopt the spatial RNN to capture the recursive matching structure, which has good performance in many semantic matching tasks. Our model has the good performance in paraphrasing identification and answer selection tasks.

Our models can be further extended in many respects. In the word interactions level based on the word identities, we just adopt the *xnor* to capture the interaction of two words. We can define the more interpretable signals to enrich the signals for the word level interaction. In the future we plan to increase this signals, then the RI-Match will be rich for the semantic matching.

Acknowledgments. This work was funded by the 973 Program of China under Grant No. 2014CB340401, the National Natural Science Foundation of China (NSFC) under Grants No. 61425016, 61472401, 61722211, 61773362, and 20180290, the Youth Innovation Promotion Association CAS under Grants No. 20144310, and 2016102, and the National Key R&D Program of China under Grants No. 2016QY02D0405.

References

1. Graves, A., Schmidhuber, J.: Offline handwriting recognition with multidimensional recurrent neural networks. In: Advances in Neural Information Processing Systems, pp. 545–552 (2009)
2. He, H., Gimpel, K., Lin, J.: Multi-perspective sentence similarity modeling with convolutional neural networks. In: Proceedings of the 2015 Conference on Empirical Methods in Natural Language Processing, pp. 1576–1586 (2015)
3. Hu, B., Lu, Z., Li, H., Chen, Q.: Convolutional neural network architectures for matching natural language sentences. In: Advances in Neural Information Processing Systems, pp. 2042–2050 (2014)
4. Huang, Z., Xu, W., Yu, K.: Bidirectional LSTM-CRF models for sequence tagging. arXiv preprint arXiv:1508.01991 (2015)
5. Kalchbrenner, N., Grefenstette, E., Blunsom, P.: A convolutional neural network for modelling sentences. arXiv preprint arXiv:1404.2188 (2014)
6. Leal-Taixé, L., Canton-Ferrer, C., Schindler, K.: Learning by tracking: siamese CNN for robust target association. In: Proceedings of the IEEE Conference on Computer Vision and Pattern Recognition Workshops, pp. 33–40 (2016)
7. Pang, L., Lan, Y., Guo, J., Xu, J., Wan, S., Cheng, X.: Text matching as image recognition. In: AAAI, pp. 2793–2799 (2016)
8. Qiu, X., Huang, X.: Convolutional neural tensor network architecture for community-based question answering. In: IJCAI, pp. 1305–1311 (2015)
9. Socher, R., Chen, D., Manning, C.D., Ng, A.: Reasoning with neural tensor networks for knowledge base completion. In: Advances in Neural Information Processing Systems, pp. 926–934 (2013)
10. Varior, R.R., Shuai, B., Lu, J., Xu, D., Wang, G.: A siamese long short-term memory architecture for human re-identification. In: Leibe, B., Matas, J., Sebe, N., Welling, M. (eds.) ECCV 2016. LNCS, vol. 9911, pp. 135–153. Springer, Cham (2016). https://doi.org/10.1007/978-3-319-46478-7_9
11. Wan, S., Lan, Y., Guo, J., Xu, J., Pang, L., Cheng, X.: A deep architecture for semantic matching with multiple positional sentence representations. In: AAAI, vol. 16, pp. 2835–2841 (2016)
12. Wan, S., Lan, Y., Xu, J., Guo, J., Pang, L., Cheng, X.: Match-SRNN: modeling the recursive matching structure with spatial RNN. arXiv preprint arXiv:1604.04378 (2016)
13. Wang, Y., Huang, M., Zhao, L., et al.: Attention-based LSTM for aspect-level sentiment classification. In: Proceedings of the 2016 Conference on Empirical Methods in Natural Language Processing, pp. 606–615 (2016)
14. Wang, Z., Hamza, W., Florian, R.: Bilateral multi-perspective matching for natural language sentences. arXiv preprint arXiv:1702.03814 (2017)
15. Wang, Z., Ittycheriah, A.: FAQ-based question answering via word alignment. arXiv preprint arXiv:1507.02628 (2015)
16. Wang, Z., Mi, H., Ittycheriah, A.: Sentence similarity learning by lexical decomposition and composition. arXiv preprint arXiv:1602.07019 (2016)
17. Yin, W., Schütze, H., Xiang, B., Zhou, B.: ABCNN: attention-based convolutional neural network for modeling sentence pairs. arXiv preprint arXiv:1512.05193 (2015)

Modeling Relations Between Profiles and Texts

Minoru Yoshida$^{(\boxtimes)}$, Kazuyuki Matsumoto, and Kenji Kita

Institute of Technology and Science, University of Tokushima,
2-1, Minami-josanjima, Tokushima 770-8506, Japan
{mino,matumoto,kita}@is.tokushima-u.ac.jp

Abstract. We propose a method to model Twitter texts and user profiles simultaneously by considering the relations between the texts and profiles to obtain the distributed representations of the words in both.

1 Introduction

The recent successes of several SNSs provide us with plenty of text data written by various people. We can find user profile data, i.e., texts that describe the users' favorite things, living places, etc., in most such SNS text data. In this paper, we propose a method for mining such user profile texts. We especially propose a method to model the relations between *user profiles* and *texts* using Twitter data as a test case.

This task is also regarded as the modeling of the users' *viewpoints*. Here a viewpoint means a way of seeing things, which can vary from user to user. For example, even though the word "Tokyo" has no ambiguity, different users in different situations use the word differently, e.g., users inside Tokyo talk about the living environment in Tokyo, whereas users outside Tokyo talk about the sightseeing spots of Tokyo. Note that this term is also known as *stances*, especially in Natural Language Processing communities. Stance classification is the classification of a group of texts that discuss a particular target concept into pros (in favor of) or cons (against,) or neutral toward the target concept. Stance classification has been an active area of research within the last decade, reflecting the fact that many users have been discussing various topics in social media within recent years.

Modeling texts according to such viewpoints contributes to more intelligent social media survey systems that, for example, filter out tweets outside of a user's interest, or conversely, propose tweets discussing the same issue from opposite viewpoints.

We propose an algorithm for viewpoint modeling that is based on word-embedding techniques. We propose an embedding method to obtain vectors that reflect not only tweets but also the *profiles of the users* associated with the tweets on the basis of the assumption that users having similar preferences also employ similar words in their profiles and tend to use similar words in tweets that discuss the same issues.

© Springer Nature Switzerland AG 2018
Y.-H. Tseng et al. (Eds.): AIRS 2018, LNCS 11292, pp. 103–109, 2018.
https://doi.org/10.1007/978-3-030-03520-4_10

2   Related Studies

Stance classification is the classification of given posts into pros and cons (or labeling each post with a pro or con label) and has been actively studied within the last decade for online debates.

Recently, stance classification on social media, especially for Twitter data, has been an active field of research. Rejadesingan et al. proposed a label propagation method for this task [2]. Du et al. proposed an attention mechanism that learns the weights for indicating how important each word is for the target [2]. Ebrahimi et al. hinge-loss MRF reflects some structures on the basis of friendly relations or tweet similarities for tweet labeling [4]. Sasaki et al. proposed a matrix decomposition method for the collaborative filtering of topics (i.e., "users supporting topic A also tend to support that topic B") [13]. Recently, SemEval2016 proposed a tweet stance classification task [10] and reported that the top system for supervised tasks used RNN, whereas the top system for semisupervised tasks used deep CNN with rule-based annotation. Other systems in SemEval include Ebrahimi et al.'s system, which models sentiments, targets, and stances simultaneously [3]. To the best of our knowledge, the current study is the first to analyze user profiles in learning word embeddings for the classification of viewpoints/stances.

There are some applications that use stance classification. Munson et al. reported a user study that employed a browser widget to encourage the reading of texts that held the opposite viewpoint [11]. Lu et al. reported user recommendation systems based on the results of stance classification [8]. Garimella et al. proposed a graph-based link recommendation algorithm [5].

Also, several studies have extracted or classified users by their profile texts to better model the texts. For example, [7] used Twitter profile texts to classify users' political affiliations into "Democrats", "Republicans" or "Unknown". However, as far as we know, our work is the first work to model user profile texts in general.

3   Problem Setting and Proposed Method

We assume that the input to the system was a list of tweet-profile pairs (t_i, p_i). t_i and p_i are decomposed into a list of words. $w \in x$ denotes the inclusions of the word w in x. Our method learns the embedding vector for each word in t_i and p_i by using t_i and p_i interactively, thereby contributing to obtaining the embedding vectors that reflect the profile-tweet co-occurrences.

We follow the implementation of word2vec, which learns the vectors using the Skip-Gram model with negative sampling (SGNS) [9].

Also, we define the objective function as the sum of the score for every (t_i, p_i) pair. And use four vectors – *tweet vectors* v_w^s, *tweet context vectors* v_w^{sc}, *profile vectors* v_w^p, and *profile context vectors* v_w^{pc} (instead of the output and input vectors in word2vec).

These four vectors are defined as follows.

Tweet vectors and tweet context vectors: We use Skip-Gram model for predicting a tweet vector for one word in a tweet from other words in the same tweet represented by tweet context vectors, which result in maximizing the value of the dot product of v_w^s and v_z^{sc} where w is the target word and z is the context word.

Profile vectors and profile context vectors: For words in the profile, profile vectors v_w^v are predicted by profile context vectors v_z^{vc}, which result in maximizing the value of the dot product of v_w^p and v_z^{pc} where w is the target word and z is the context word.

We can also consider another relation between the profile vectors and tweet vectors that reflects the co-occurrence relations between profile words and tweet words, as well as results in maximizing the value of the dot product of v_w^p and v_z^{sc} where w is the word in the profile and z is the word in the tweet. This additional objective is intended to predict the word in the profile p_i from the tweet context vector in the tweet t_i. Our intuition for using this relation is the similarity between the preferences of users who tend to use similar vocabularies because they are interested in similar issues.

For each pair (t_i, p_i), we take every pair of the target word w and its context word c. A vector for w and a vector for c are selected from "tweet", "tweet context", "profile", or "profile context" according to the position of each word. More details are given in the following sections. The parameters are learned to maximize the likelihood of the training samples. The definition of the likelihood function is shown below.

The objective function to maximize for word w in the profile and word z in the tweet/profile is:

$$
l_{(z,w)} = \begin{cases} \log \sigma(v_z^{sc} \cdot v_w^p) + \sum_{k=1}^{K} \log \sigma(-v_z^{sc} \cdot v_{w_k}^p) \\ \qquad \text{(if z is in the tweet)} \\ \log \sigma(v_z^{pc} \cdot v_w^p) + \sum_{k=1}^{K} \log \sigma(-v_z^{pc} \cdot v_{w_k}^p) \\ \qquad \text{(if z is in the profile).} \end{cases}
$$

where K is the number of negative samples and σ is the sigmoid function. Negative samples w_k are randomly selected at each learning step.

The final objective function for each pair (t_i, p_i) is the sum of the above functions for all (w, t) pairs:

$$
l = \sum_{w \in p_i} [\sum_{z \in p_i, z \neq w} \{\log \sigma(v_z^{pc} \cdot v_w^p) + \sum_{k=1}^{K} \log \sigma(-v_z^{pc} \cdot v_{w_k}^p)\}
$$

$$
+ [\sum_{z \in t_i} \{\log \sigma(v_z^{sc} \cdot v_w^p) + \sum_{k=1}^{K} \log \sigma(-v_z^{sc} \cdot v_{w_k}^p)\}]
$$

$$
+ \sum_{w \in t_i} [\sum_{z \in t_i, z \neq w} \{\log \sigma(v_z^{sc} \cdot v_w^s) + \sum_{k=1}^{K} \log \sigma(-v_z^{sc} \cdot v_{w_k}^s)\}]
$$

Note that the profile vectors are combined with both the profile context vectors and tweet context vectors in the objective function[1].

3.1 Parameter Learning

We used Stochastic Gradient Descent (SGD) in the same way as the word2vec implementation [6] by taking each pair of a word and context as one training datum and optimizing on this datum. For every pair in the training data, this process was iterated.[2] The number of iterations was set to 10.

4 Experiments

We collected data from Japanese tweets (and profiles) by using a Twitter streaming API (random sampling) from July to October 2017. We assumed that the tweets containing "RT @" or "http" were informative and suitable for recommendations and retained only such tweets.[3] Retweet indicators (i.e., "RT @(username)") and URLs were removed from each tweet.

We obtained embedding vectors for each word with frequencies over the threshold (currently ≥ 10). We set the number of dimensions to 50. We tested our four models and the baseline model that used vector space models with one-hot vector representations.

For evaluation, we used the *profile ranking task* and *tweet ranking task*. In the profile ranking task, we were given a tweet and a list of profile texts. We ranked the profile texts according to the relatedness to the given tweet. We collected the tweets retweeted by two or more users, and a set of the users as the correct answers for the tweets. We calculated the *average precisions* [1] of the list of profiles for each tweet. This is the value defined in the ranked list of elements where some elements are *correct* (*i.e.*, included in some *set of answers*) and others are not. Given a list of extracted terms $\langle c_1, c_2, \cdots c_n \rangle$ ranked by the score, and a set of answers $S = \{s_1, s_2, \cdots\}$, the average precision of the result list was calculated as $\frac{1}{|S|} \sum_{1 \leq k \leq n} r_k \cdot precision(k)$, where $precision(k)$ is the accuracy (i.e., the ratio of correct answers to all answers) of the top k candidates, and r_k represents whether the k-th document is relevant (1) or not (0) (*i.e.*, $r_k = 1$ if $c_k \in S$, and $r_k = 0$ otherwise). We collected all (tweet, set of profiles) pairs with sizes (*i.e.*, the size of "set of profiles") of 10 or more. All the profiles in these pairs were merged into one set of profiles, which were then ranked.

Conversely, in the tweet ranking task, we were given a user profile text and a list of tweets. We ranked the texts according to the relatedness to the given

---

[1] Also note that only two of these six terms are used for each (w, z) pair, which makes the SGD implementation for this model almost the same as that of word2vec.

[2] We parallelized this by the Hogwild approach [12], by assigning each process a subset of tweet-profile pairs in the corpus.

[3] We observed that the removing this constraint did not affect the results so much because of the large proportion (over 65% for both of the query "Hoikuen" and "Tokushima") of the tweets contained "RT" or "http".

user profile. The answer for each user was the list of tweets posted by the user. We collected all (profile, set of tweets) pairs with sizes (*i.e.,* the size of "set of tweets") of 3 or more. All the tweets in these pairs were merged into one set of tweets, which were then ranked.

We associated one vector, which was calculated as the average of the (context) sentence vectors of the words in a tweet, to each tweet in the test set.

We selected two keywords "Hoikuen" (nursery) and "Tokushima" (the name of an area in Japan). The former is a word that has several viewpoints, such as political, childcare, recruitment, etc. (discussed in more details in the section on the experimental results). The latter is an example of a regional word that will change its meaning according to whether users live in that area or not. Tweets including each keyword were collected. Then, we automatically constructed the test data by selecting users with two or more tweets and selected half of such users as the test data. We excluded tweets similar to one of the tweets in the test data[4].

We compared two types of implementations of our methods and two types of baselines. "SEPARATE" means that the words in the profiles and those in the tweets were treated as different words even if the profiles and tweets shared the same vocabulary. "FLAT" conversely assigned the same vector to the same word regardless whether the word was in the profiles or tweets. "FLAT" can be regarded as a simple SGNS model applied to the texts obtained by concatenating profile and tweets into one text. "BASELINE" gave a random vector to each word. "RANDOM" simply ranked the profiles or tweets randomly.

We obtained 7,493 tweets and 140 test users for the query "Hoikuen", and 8,401 tweets and 290 users for the query "Tokushima". The numbers of test tweets for the profile prediction task was 13 for "Tokushima" and 37 for "Hoikuen". The numbers of test profiles for the tweet prediction task was 87 for "Tokushima" and 25 for "Hoikuen". Tweets collected by the query "Hoikuen" included the advertisements of nurseries, tweets by children's nurses, tweets by parents, tweets discussing public policies, such as budgets for nurseries, and job recruitment by nurseries. Tweets collected by the query "Tokushima" included the soccer team in Tokushima, restaurants and sightseeing spots in Tokushima, travel reports in Tokushima, advertisement events for a Tokushima-related product, and recruitment by businesses in Tokushima.

Table 1 shows the results for the all algorithms. Each result is the average of 3 trials. We observed that in most cases the proposed modeling produced better results than did the baselines except for the tweet prediction task for the query "Hoikuen", for which a few users who shared exactly the same vocabulary in their profiles and tweets (e.g., a user who had one author name in his profile and frequently tweeted about that author's books) contributed to decreasing the effect of learning the word semantic vectors. "SEPARATE" performed better for the profile prediction task, whereas "FLAT" performed better for the tweet prediction task.

---

[4] We calculated the length of the overlap string between two tweets. If the overlap length exceeded the half of the length of the (longer) tweet, we regarded them similar.

Table 1. Results in average precision (%) for both tasks

Query	Flat	Separate	Baseline	Random
Profile prediction task				
Tokushima	11.78	**13.97**	10.678	9.814
Hoikuen	4.464	**4.898**	3.683	3.331
Tweet prediction task				
Tokushima	**8.240**	5.995	5.549	2.031
Hoikuen	8.138	8.035	**13.24**	6.680

5   Conclusions and Future Work

We proposed a method to model profiles and texts simultaneously. Our method learns the word distributed representations found in Twitter profiles and tweets by modeling the relations between the words in the user profiles and those in the tweets. Future work would include the experiments with more queries and an investigation into which models are suitable for which queries.

Acknowledgement. This work was supported by JSPS KAKENHI Grant Numbers JP18K11549, JP15K00309, JP15K00425, JP15K16077.

References

1. Chakrabarti, S.: Mining the Web: Discovering Knowledge from Hypertext Data. Morgan-Kaufmann Publishers, Burlington (2002)
2. Du, J., Xu, R., He, Y., Gui, L.: Stance classification with target-specific neural attention. In: Proceedings of IJCAI 2017, pp. 3988–3994 (2017)
3. Ebrahimi, J., Dou, D., Lowd, D.: A joint sentiment-target-stance model for stance classification in tweets. In: Proceedings of COLING 2016, pp. 2656–2665 (2016)
4. Ebrahimi, J., Dou, D., Lowd: D.: Weakly supervised tweet stance classification by relational bootstrapping. In: Proceedings of EMNLP 2016, pp. 1012–1017 (2016)
5. Garimella, K., Morales, G.D.F., Gionis, A., Mathioudakis, M.: Reducing controversy by connecting opposing views. In: Proceedings of WSDM 2017, pp. 81–90 (2017)
6. Ji, S., Satish, N., Li, S., Dubey, P.: Parallelizing Word2Vec in shared and distributed memory. CoRR abs/1604.04661 (2016)
7. Joshi, A., Bhattacharyya, P., Carman, M.J.: Political issue extraction model: a novel hierarchical topic model that uses tweets by political and non-political authors. In: WASSA, NAACL-HLT 2016, pp. 82–90 (2016)
8. Lu, H., Caverlee, J., Niu, W.: BiasWatch: a lightweight system for discovering and tracking topic-sensitive opinion bias in social media. In: Proceedings of CIKM 2015, pp. 213–222 (2015)
9. Mikolov, T., Sutskever, I., Chen, K., Corrado, G.S., Dean, J.: Distributed representations of words and phrases and their compositionality. In: Proceedings of NIPS 2013, pp. 3111–3119 (2013)

10. Mohammad, S., Kiritchenko, S., Sobhani, P., Zhu, X.D., Cherry, C.: SemEval-2016 task 6: detecting stance in tweets. In: Proceedings of SemEval@NAACL-HLT 2016, pp. 31–41 (2016)
11. Munson, S.A., Lee, S.Y., Resnick, P.: Encouraging reading of diverse political viewpoints with a browser widget. In: Proceedings of ICWSM 2013, pp. 419–428 (2013)
12. Recht, B., Re, C., Wright, S.J., Niu, F.: HOGWILD: a lock-free approach to parallelizing stochastic gradient descent. In: Proceedings of NIPS 2011, pp. 693–701 (2011)
13. Sasaki, A., Hanawa, K., Okazaki, N., Inui, K.: Other topics you may also agree or disagree: modeling inter-topic preferences using tweets and matrix factorization. In: Proceedings of ACL 2017, no. 1, pp. 398–408 (2017)

Recommendation and Classification

Missing Data Modeling with User Activity and Item Popularity in Recommendation

Chong Chen, Min Zhang$^{(\boxtimes)}$, Yiqun Liu, and Shaoping Ma

Department of Computer Science and Technology, Institute for Artificial Intelligence,
Beijing National Research Center for Information Science and Technology,
Tsinghua University, Beijing 100084, China
z-m@tsinghua.edu.cn

Abstract. User feedback such as movie watching history, ratings and consumptions of products, is valuable for improving the performance of recommender systems. However, only a few interactions between users and items can be observed in implicit data. The missing of a user-item entry is caused by two reasons: the user didn't see the item (in most cases); or the user saw but disliked it. Separating these two cases leads to modeling missing interactions at a finer granularity, which is helpful in understanding users' preferences more accurately. However, the former case has not been well-studied in previous work. Most existing studies resort to assign a uniform weight to the missing data, while such a uniform assumption is invalid in real-world settings. In this paper, we propose a novel approach to weight the missing data based on user activity and item popularity, which is more effective and flexible than the uniform-weight assumption. Experimental results based on 2 real-world datasets (Movielens, Flixster) show that our approach outperforms 3 state-of-the-art models including BPR, WMF, and ExpoMF.

Keywords: Recommender systems · Collaborative filtering
Matrix factorization · Implicit feedback

1 Introduction

In the era of information explosion, not only are users not easy to find items they are interested in, such as news, merchandise, music, etc., it is also difficult for providers to display products accurately to the target population. In this case, the role of recommender systems is becoming increasingly important.

The key to recommender systems is to infer users' preferences from historical records, such as ratings, reviews, clicks, and consumptions, etc. Compared to explicit feedback (e.g., ratings and reviews), implicit feedback like users' video viewing and product purchase history, doesn't require users' extra operations and can be tracked automatically. Therefore it is much easier for providers to collect. However, implicit feedback is more challenging to utilize, since it is binary

© Springer Nature Switzerland AG 2018
Y.-H. Tseng et al. (Eds.): AIRS 2018, LNCS 11292, pp. 113–125, 2018.
https://doi.org/10.1007/978-3-030-03520-4_11

and only has positive examples. When inferring users' preferences, the items without interactions are essential. These items are referred to as missing data in recommender systems.

Previous studies [6,7,12,15] deal with this problem in two ways: either randomly sampling negative instances from the missing data, or treating all of them as negative. However, an important fact cannot be overlooked when we revisit this problem—many items, actually a large number of them were not clicked because the user didn't see them, rather than disliked them. If an item was never noticed by the user, then no consumption can possibly be made, and the missing interaction implies no particular positive or negative preference of the user at all. Many previous approaches have not distinguished these two cases. They assign a uniform weight to the missing data, assuming that the missing entries have equal probability to be negative feedback, and hence we prefer that these studies are biased in terms of modeling users' preferences accurately.

Compared with the previous work, [6,10] propose to weight the missing data based on item popularity. In the same condition, popular items are more likely to be known by users in general [4], and thus it is reasonable to think that a missing popular item is more probable to be truly not attracted (as opposed to unknown) to the user. Using item popularity to model missing data is effective, but it still has a flaw: not considering the differences among different users and making all users have the same weight to a missing item.

In fact, users with different degrees of activity usually have different visibility for items. Inactive users (or new users) tend to browse popular items, while active users are more likely to browse unpopular items. In this paper, we define user activity as the total number of items clicked by the user, while item popularity is defined as the number of users who clicked on it. In Fig. 1, we show the relationship between user activity and item popularity on Flixster dataset. As we can see from the figure, the curve shows an obvious downward trend, which indicates that unpopular items are more likely to be known by active users.

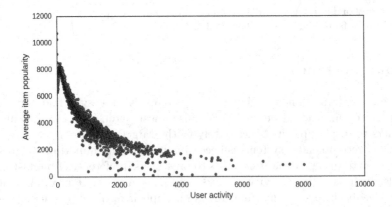

Fig. 1. The relationship between user activity and item popularity on Flixster dataset

In this paper, we propose a new method named UIMF, considering both user activity and item popularity to weight missing data and make recommendation for implicit feedback. Experiments have been conducted on 2 real-world datasets in comparison with 3 state-of-art approaches. The encouraging results verify the effectiveness of our model.

The remainder of the paper is organized as follows. The next section introduces relevant prior work on collaborative filtering and implicit feedback. Section 3 gives a detailed introduction about modeling missing data based on user activity and item popularity. We conduct experiments and the results are presented in Sect. 4. Finally, we conclude the paper in Sect. 5.

2   Related Work

User feedback is frequently seen in real-life scenarios and is usually in different forms, such as ratings, reviews, clicks and consumptions of items. Handling user feedback has been a key issue in recommender systems. Many studies have been made to enhance the performance of recommendation through these historical records.

In recent years, matrix factorization (MF) has become the most popular collaborative filtering approach [9,16]. The original MF models were designed to model users' explicit feedback by mapping users and items to a latent factor space, such that user-item relationships (e.g., ratings) can be captured by their latent factors' dot product. Based on that, many research efforts have been devoted to enhancing MF, such as integrating it with neighbor-based models [8] and extending it to factorization machines [14] for a generic modeling of features. However, it is still problematic to apply traditional matrix factorization to implicit feedback due to the lack of negative instances.

To this end, two basic strategies have been proposed in previous studies [6]: sample based learning that samples negative instances from the missing data [12, 15] and whole-data based learning that sees all the missing data as negative [7, 11,18]. Compared with sample based methods, whole-data based methods can model the full data with a potentially higher coverage, and thus may achieve a better performance if the parameters are set properly.

Most existing whole-data based methods [3,7,11,13,17,18] assign a uniform weight to all the missing data, assuming that the missing entries have the same probability to be negative feedback, which facilitates the efficiency of the algorithm, but limits the flexibility and extensibility of the model. As discussed in Sect. 1, [6,10] are the only works that consider item popularity for weighting missing feedback. [10] proposes the concept of exposure to model whether an item has been observed by a user. It thinks that popular items are more likely to be known by users and thus gives a higher weight to popular items. [6] devises a new object function of matrix factorize, in which item popularity is used to model the confidence that item i missed by users is a truly negative instance. Unlike the previous two methods, [15] adopts popularity-based oversampling for learning BPR, which basically samples popular items as negative feedback with a higher probability.

In our work, we propose a novel approach to model missing data and utilize Bayesian approaches to estimate the weight of them. Differing from previous studies, our work uses both user activity and item popularity in modeling missing data, while only item popularity is considered in [6,10,15].

To our knowledge, this work is the first attempt to exploit user activity for modeling missing data.

3   Missing Data Modeling

In this section, we present our model (UIMF). In Sect. 3.1, we briefly introduce the model of modeling missing data with user activity and item popularity. In Sect. 3.2, we derive inference procedures for our model. The connections between our approach and other models are shown in Sect. 3.3. The variables of our model are listed in Table 1.

Table 1. Variables introduction

Variables	Meaning
y_{ui}	User-Item Interaction: i.e. whether user u has clicked on item i
θ_u	The preference latent factor vector of user u
β_i	The attribute latent factor vector of item i
a_{ui}	Whether user u has seen item i
p_{ui}	The probability that user u sees item i
μ_i	The parameter of "item popularity only" strategy
η_u	The parameter of "user activity only" strategy
ω_{ui}	The parameter of "both user activity and item popularity" strategy
(α_1, α_2)	The parameter of Beta distribution

3.1   Model Description

As implicit data is very sparse, the interactions between users and items that can be observed are rather limited. The variable y_{ui} indicates whether there is an interaction between user u and item i (if there is an interaction, $y_{ui} = 1$, otherwise $y_{ui} = 0$). The general idea of this model is that, many items were not clicked or consumed not because the user didn't like them, but because the user didn't see them. When inferring users' preferences, we need to assign an appropriate weight to each missing entry according to the probability that the user sees the item. We use a_{ui} to indicate whether user u has seen item i ($a_{ui} = 1$ means that u has seen i, and otherwise $a_{ui} = 0$). Then the variable p_{ui} is introduced to capture the probability that $a_{ui} = 1$. If p_{ui} is large, very possible that user u has seen item i but choose not to click on it, then the confidence that i is a truly negative instance should also be large (The converse argument also holds for low values of p_{ui}).

The value of p_{ui} is related to the popularity of item i and the activity of user u. If an item is popular, then it is more likely to be seen by users. Similarly, if a user is active, then the probabilities that he sees items are higher. Therefore, the weight of the missing entry (y_{ui}, $y_{ui} = 0$) should be assigned large if u is active and i is popular.

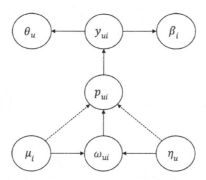

Fig. 2. Graphical representation of our MF model. A directed edge from node a to node b denotes that the variable b depends on the value of variable a. μ_i and η_u are derived from item popularity and user activity respectively. ω_{ui} is derived from μ_i and η_u and uses both user activity and item popularity.

Based on the above idea, we propose a new method for matrix factorization. The graphical model is presented in Fig. 2. Given the condition that user u has seen item i ($a_{ui} = 1$), the probability that user u would click on item i follows Gaussian distribution [16] (λ_y is precision for corresponding Gaussian distribution):

$$y_{ui}|(a_{ui} = 1) \sim N\left(\theta_u^T \beta_i, \lambda_y^{-1}\right), \forall u, i \tag{1}$$

and the variable a_{ui} follows Bernoulli distribution:

$$a_{ui} \sim Bernoulli(X), \forall u, i \tag{2}$$

where X can be replaced by μ_i, η_u and ω_{ui}, which represent the priors of the Bernoulli distribution derived from "item popularity only", "user activity only" and "both user activity and item popularity".

In Fig. 3, we show an example of using different strategies to weight missing data. The strategy of "item popularity only" makes every user has the same weight to a missing item, while "user activity only" makes every missing item has no difference for a user. These two methods do not correspond with the actual situation. Different from them, in our model we propose to weight every missing entry individually by considering both user activity and item popularity, which is more practical in real-word settings.

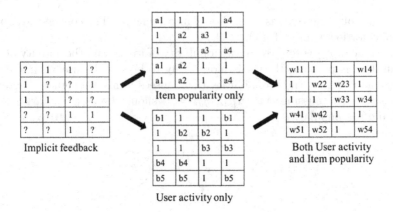

Fig. 3. Different strategies to weight missing data. a_i, b_u and w_{ui} denote the weight of missing entry and we use $W = [w_{ui}]_{M \times N}$ to represent the weight matrix.

However, one thing can not be ignored: the high space complexity makes it impossible to explicitly store the weight matrix W even for medium-sized datasets. As an alternative, we firstly capture μ_i and η_u from Beta distributions respectively:

$$\mu_i \sim Beta(\alpha_1, \alpha_2), \forall i; \eta_u \sim Beta(\alpha_1', \alpha_2'), \forall u \qquad (3)$$

After that, we construct the necessary part of ω_{ui} when it is used to update the user/item factors by adding these two variables with individual weights.

Our proposed model is named UIMF, as it considers both user activity and item popularity for modeling missing feedback.

3.2 Model Learning

We use expectation-maximization (EM) [2] algorithm to infer the parameters of UIMF.

The E step that can be derived from Bayesian Theorem is:

$$p_{ui}(y_{ui} = 0) = \frac{\omega_{ui} \cdot N\left(0|\theta_u^T \beta_i, \lambda_y^{-1}\right)}{\omega_{ui} \cdot N\left(0|\theta_u^T \beta_i, \lambda_y^{-1}\right) + (1 - \omega_{ui})} \qquad (4)$$

where $N\left(0|\theta_u^T \beta_i, \lambda_y^{-1}\right)$ stands for the probability density function of $N\left(\theta_u^T \beta_i, \lambda_y^{-1}\right)$ evaluated at 0. Since p_{ui} indicates the probability that user u sees item i, we can define $p_{ui}(y_{ui} = 1) = 1$.

Therefore, we can estimate the user/item factors in the following M step, which can be derived based on Alternative Least Square (ALS) optimization (λ_s is corresponding precision matrix for Gaussian distribution; I_k is the identity matrix.):

$$\theta_u \leftarrow \left(\lambda_y \sum_i p_{ui}\beta_i\beta_i^T + \lambda_\theta I_k\right)^{-1} \left(\sum_i \lambda_y p_{ui}y_{ui}\beta_i\right) \qquad (5)$$

$$\beta_i \leftarrow \left(\lambda_y \sum_u p_{ui} \theta_u \theta_u^T + \lambda_\theta I_k \right)^{-1} \left(\sum_u \lambda_y p_{ui} y_{ui} \theta_u \right) \tag{6}$$

We update μ_i, η_u and ω_{ui} as follows:

- **Item popularity only:**

$$\mu_i \leftarrow \frac{\alpha_1 + \sum_u p_{ui} - 1}{\alpha_1 + \alpha_2 + \|U\| - 2} \tag{7}$$

- **User activity only:**

$$\eta_u \leftarrow \frac{\alpha_1' + \sum_i p_{ui} - 1}{\alpha_1' + \alpha_2' + \|I\| - 2} \tag{8}$$

- **Both user activity and item popularity:**

$$\omega_{ui} \leftarrow k \cdot \mu_i + (1 - k) \cdot \eta_u \tag{9}$$

To make recommendation, we consider both the probability that the user sees items and his preference (an item is more likely to be seen means more likely to be clicked). Therefore, we construct the following ranking score:

$$\bar{y_{ui}} = p_{ui} * \theta_u^T \beta_i \sim \omega_{ui} * \theta_u^T \beta_i \tag{10}$$

and the items (unclicked/ unconsumed) are ranked in descending order $\bar{y_{ui}}$ to provide the Top-N item recommendation list.

3.3 Model Flexibility

Our proposed model can be easily converted to other models by changing the update method of ω_{ui}.

When the value of ω_{ui} is fixed to 1, we recover traditional matrix factorization [16]. When it is fixed between 0 and 1, we can get weighted matrix factorization (WMF) [7]. ExpoMF [10] is also a special case of our model which can be obtained by fixing the value of k to 1 in Eq. (9) when updating ω_{ui} and using "item popularity only" strategy.

4 Experiment

We begin by introducing the experimental settings. Then we present the experimental results conducted on 2 real-world datasets, followed by an exploratory analysis on the influence of user activity and item popularity.

4.1   Experimental Settings

Datasets. We evaluate on 2 real-world datasets: MovieLens[1] and Flixster[2]. Movielens has been widely used to evaluate collaborative filtering algorithms, the version we used contains about one million ratings. Flixster is a dataset for evaluating social information based recommendation, it has a huge amount of interactions from around one hundred thousand users and ten thousand items. The datasets have been preprocessed so that all the items have at least 5 ratings. We treat the corresponding rating as 1 as long as there is a user-item interaction, which is the same procedure adopted in many previous studies including [5–7,10,15]. The statistical details of the 2 datasets are presented in Table 2.

Table 2. Statistics of the evaluation datasets.

Datasets	#users	#items	#interactions	#density
MovieLens	6,040	3,706	1,000,209	4.47%
Flixster	147,229	17,318	8,093,735	0.317%

Baselines and Our Methods. We compare with the following recommendation methods for implicit feedback:

- **BPR** [15]: This is a sample based method that optimizes the pair-wise ranking between the positive and negative samples.
- **WMF** [7]: This is a whole-data based method that treats all missing interactions as negative instances and weights them uniformly.
- **ExpoMF** [10]: This is a state-of-the-art whole-data based method for item recommendation. It also treats all missing interactions as negative instances but weights them non-uniformly by item popularity.

We also compare the performance of 3 missing data modeling strategies in our model. Because the strategy of "item popularity only" is the same as ExpoMF, we present the following 2 methods:

- **UIMF**:This is the model we proposed in this paper, which takes both user activity and item popularity into consideration to weight missing interactions.
- **UMF**: This is a special case of our model, which uses the strategy of "user activity only" to weight missing interactions.

Evaluation Metrics. We adopt frequently used metrics [1] to evaluate the performance, including Recall@K, MAP@K and NDCG@K. where $rel_i = 1/0$ indicates whether the item at rank i in the Top-N list is in the testing set. For

---

[1] http://grouplens.org/datasets/movielens/1m/.
[2] http://www.sfu.ca/~sja25/datasets/.

each user, these metrics can be computed as follows (each metric is the average for all users, and MAP is the average of all AP of users).

$$Recall@K = \frac{\sum_{i=1}^{K} rel_i}{min\left(K, y_u^{test}\right)}; AP@K = \sum_{n=1}^{K} \frac{\frac{\sum_{i=1}^{n}}{n} \times rel_n}{min\left(K, y_u^{test}\right)}$$

$$DCG@K = \sum_{i=1}^{K} \frac{2^{rel_i} - 1}{log_2\left(i+1\right)}; NDCG@K = \frac{DCG@K}{IDCG@K} \tag{11}$$

To evaluate on different recommendation lengths, we set K = 10, 50 and 100 in our experiments.

Table 3. Performance comparison on 2 datasets. The best performing method is bold-faced, and the last column shows the improvements of UIMF compared to the best results in baselines and UMF. The improvements with "*" are significant with p-value <0.05, and the improvements with "**" are significant with p-value <0.01

MovieLens	BPR	WMF	ExpoMF	UMF	UIMF	UIMF vs. best
Recall@10	0.1442	0.4241	0.4250	0.4259	**0.4267**	0.18%
Recall@50	0.3876	0.4704	0.4734	0.4720	**0.4753****	0.40%
Recall@100	0.5340	0.5686	0.5723	0.5700	**0.5743****	0.34%
NDCG@10	0.4190	0.4473	0.4481	0.4495	**0.4502**	0.15%
NDCG@50	0.4071	0.4343	0.4363	0.4361	**0.4383****	0.47%
NDCG@100	0.4462	0.4721	0.4745	0.4739	**0.4766****	0.44%
MAP@10	0.2864	0.3076	0.3087	0.3100	**0.3102**	0.06%
MAP@50	0.2077	0.2290	0.2305	0.2307	**0.2320****	0.64%
MAP@100	0.2082	0.2310	0.2327	0.2327	**0.2344****	0.72%
Flixster	BPR	WMF	ExpoMF	UMF	UIMF	UIMF vs. best
Recall@10	0.1622	0.3766	0.3788	0.3826	**0.3865***	1.02%
Recall@50	0.3337	0.4909	0.4947	0.4975	**0.5029****	1.09%
Recall@100	0.4366	0.5596	0.5684	0.5659	**0.5727****	1.20%
NDCG@10	0.1227	0.3266	0.3248	0.3302	**0.3323***	0.64%
NDCG@50	0.1765	0.3534	0.3543	0.3594	**0.3638****	1.22%
NDCG@100	0.2061	0.3725	0.3760	0.3792	**0.3845****	1.40%
MAP@10	0.0893	0.2489	0.2484	0.2511	**0.2529***	0.72%
MAP@50	0.1005	0.2355	0.2379	0.2408	**0.2453****	1.87%
MAP@100	0.1046	0.2351	0.2391	0.2441	**0.2466****	1.02%

Experiments Details. We adopt the open source implementation in librec[3] to obtain the predictions of BPR; for WMF we use the open source code from

---

[3] https://www.librec.net.

github[4] since it can get a better performance than librec; and for ExpoMF, we use the source code[5] released by the authors. The parameters for baseline methods are initialized as in the corresponding paper and they are further carefully tuned around to achieve the best performance. The dimensions of latent factor vectors are set to 50 for both MovieLens and Flixster.

4.2   Performance Comparison

We perform a four-fold cross-validation in our experiments. Three folds are used for training and the rest fold is used for testing. For every dataset, we conduct all methods eight times and the average result is presented in Table 3.

We make the following observations:

First, as shown in Table 3, The methods weighting missing data non-uniformly (ExpoMF, UMF, UIMF) generally outperform the uniform weighting method WMF. We believe the benefits mainly come from the non-uniform setting of weight that derived from user activity or item popularity since it is more practical in real-world.

Secondly, our method (UIMF) using both user activity and item popularity outperforms ExpoMF and UMF, which weight missing data only by item popularity or user activity. This is because that UIMF weights each missing entry individually so that it can better capture users' preferences and items' attributes as described in Sect. 3.1. We also observe that UIMF tends to show more obvious improvement on Flixster than MovieLens. We think the reason is that Flixster is sparser and thus has more missing entries than MovieLens, which can be utilized better by UIMF since it is designed for dealing with the problem of missing data.

Another observation is that UIMF shows no significant improvement ($p<0.05$) on MovieLens dataset when the value of N is set to 10 in Top-N recommendation. However, as the value of N is set higher, the improvement of UIMF becomes more obvious, showing that our model is more accurate when it comes to covering a wider range of users' interests.

4.3   Impact of User Activity and Item Popularity

The value of the variable k in Eq. 9 determines the weights of user activity and item popularity when modeling missing data. To explore the impact of these two parts, we conduct experiments on different values of k in Eq. 9 when updating ω_{ui}. We alter the value of k with a stepsize of 0.1, and compare the performances correspondingly. Note that a value of 1 corresponds to ExpoMF and a value of 0 corresponds to UMF. The results of Recall@50, NDCG@50 and MAP@50 on MovieLens dataset are presented in Fig. 4.

From the figure above, we can first see that UIMF generally outperforms the baseline method WMF. Secondly, compared with the strategies of "user activity only" ($k = 0$) and "item popularity only" ($k = 1$), using both of them (no

---

[4] https://github.com/benanne/wmf.
[5] https://github.com/dawenl/expo-mf.

matter how much the value of k is adopted from 0 to 1) can always get a better performance, indicating the effectiveness of using both user activity and item popularity for missing data modeling.

What's more, the performances increase with the increase of variable k from 0 to 0.6, and then reach the optimal performance at around 0.6, after that, the performances gradually decline with the value of k increases from 0.8 to 1. This illustrates that user activity and item popularity may not be equally impactful when weighting missing data, and a proper value of k is needed to better combine them so that UIMF can achieve the best performance.

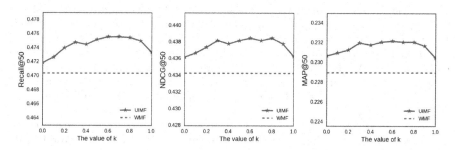

Fig. 4. The recommendation performances with different values of k on MovieLens dataset.

5   Conclusion

In this paper, we study the problem of how to model missing data in recommendation. Different from previous work that applied a uniform weight on missing interactions or just weighted them based on item popularity, we propose to consider both user activity and item popularity to weight missing data. A novel unified model (UIMF) is designed based on this idea. The major contributions of this work are:

First, we propose to consider both user activity and item popularity to model the missing data, which helps to capture users' preferences and items' attributes more accurately. As far as we know, this work is the first attempt to exploit the impact of user activity for implicit data in the literature.

Second, we design a novel unified model (UIMF) to connect both user activity and item popularity and use the variable k to control the influence of each part.

Third, extensive experiments have been conducted on 2 real-world datasets in comparison with 3 previous methods. Statistically our proposed model (UIMF) achieves significantly better performances in most cases, which verifies the effectiveness of the model.

In the future, we will make further improvements to the model to address the problem of high time complexity due to the non-uniform weighting of missing data.

Acknowledgments. We thank the anonymous reviewers for their valuable comments and suggestions. This work is supported by the Natural Science Foundation of China under Grant No.: 61672311 and 61532011.

References

1. Cremonesi, P., Koren, Y., Turrin, R.: Performance of recommender algorithms on top-N recommendation tasks. In: Proceedings of the Fourth ACM Conference on Recommender Systems, pp. 39–46. ACM (2010)
2. Dempster, A.P., Laird, N.M., Rubin, D.B.: Maximum likelihood from incomplete data via the EM algorithm. J. Roy. Stat. Soc. Ser. B (Methodol.) **39**, 1–38 (1977)
3. Devooght, R., Kourtellis, N., Mantrach, A.: Dynamic matrix factorization with priors on unknown values. In: Proceedings of the 21th ACM SIGKDD International Conference on Knowledge Discovery and Data Mining, pp. 189–198. ACM (2015)
4. He, X., Gao, M., Kan, M.Y., Liu, Y., Sugiyama, K.: Predicting the popularity of web 2.0 items based on user comments. In: Proceedings of the 37th International ACM SIGIR Conference on Research & Development in Information Retrieval, pp. 233–242. ACM (2014)
5. He, X., Liao, L., Zhang, H., Nie, L., Hu, X., Chua, T.S.: Neural collaborative filtering. In: Proceedings of the 26th International Conference on World Wide Web, pp. 173–182. International World Wide Web Conferences Steering Committee (2017)
6. He, X., Zhang, H., Kan, M.Y., Chua, T.S.: Fast matrix factorization for online recommendation with implicit feedback. In: Proceedings of the 39th International ACM SIGIR Conference on Research and Development in Information Retrieval, pp. 549–558. ACM (2016)
7. Hu, Y., Koren, Y., Volinsky, C.: Collaborative filtering for implicit feedback datasets. In: Eighth IEEE International Conference on Data Mining, 2008. ICDM 2008, pp. 263–272. IEEE (2008)
8. Koren, Y.: Factorization meets the neighborhood: a multifaceted collaborative filtering model. In: Proceedings of the 14th ACM SIGKDD International Conference on Knowledge Discovery and Data Mining, pp. 426–434. ACM (2008)
9. Koren, Y., Bell, R., Volinsky, C.: Matrix factorization techniques for recommender systems. Computer **42**(8) (2009)
10. Liang, D., Charlin, L., McInerney, J., Blei, D.M.: Modeling user exposure in recommendation. In: Proceedings of the 25th International Conference on World Wide Web, pp. 951–961. International World Wide Web Conferences Steering Committee (2016)
11. Ning, X., Karypis, G.: SLIM: sparse linear methods for top-n recommender systems. In: 2011 IEEE 11th International Conference on Data Mining (ICDM), pp. 497–506. IEEE (2011)
12. Pan, R., et al.: One-class collaborative filtering. In: Eighth IEEE International Conference on Data Mining. ICDM 2008, pp. 502–511. IEEE (2008)
13. Pilászy, I., Zibriczky, D., Tikk, D.: Fast als-based matrix factorization for explicit and implicit feedback datasets. In: Proceedings of the Fourth ACM Conference on Recommender Systems, pp. 71–78. ACM (2010)
14. Rendle, S.: Factorization machines. In: 2010 IEEE 10th International Conference on Data Mining (ICDM), pp. 995–1000. IEEE (2010)

15. Rendle, S., Freudenthaler, C., Gantner, Z., Schmidt-Thieme, L.: BPR: Bayesian personalized ranking from implicit feedback. In: Proceedings of the Twenty-Fifth Conference on Uncertainty in Artificial Intelligence, pp. 452–461. AUAI Press (2009)
16. Salakhutdinov, R., Mnih, A.: Probabilistic matrix factorization. In: NIPS, vol. 1, pp. 1–2 (2007)
17. Steck, H.: Training and testing of recommender systems on data missing not at random. In: Proceedings of the 16th ACM SIGKDD International Conference on Knowledge Discovery and Data Mining, pp. 713–722. ACM (2010)
18. Volkovs, M., Yu, G.W.: Effective latent models for binary feedback in recommender systems. In: Proceedings of the 38th International ACM SIGIR Conference on Research and Development in Information Retrieval, pp. 313–322. ACM (2015)

Influence of Data-Derived Individualities on Persuasive Recommendation

Masashi Inoue[1](✉) [iD] and Hiroshi Ueno[2]

[1] Tohoku Institute of Technology, Yagiyama Kasumicho 35-1, Sendai, Japan
m.inoue@acm.org
http://www.ice.tohtech.ac.jp/~inoue/index.html
[2] Yamagata University, Jyonan 4-3-16, Yonezawa, Japan

Abstract. In this study, two machine learning based approaches have been compared that can add personal communication traits to a conversational recommender system. The first approach involves the creation of generative models for reactive tokens such as backchannels. The second approach involves a method for rewriting the conversational text by applying machine translation. Both approaches can impart personal communication traits to systems that incorporate a dialogue corpus. Two methods were implemented for a persuasive recommender system and their positive or negative effects based on an individual's personality were experimentally analyzed through a restaurant ranking task. The results suggest that addition of personal communication traits decrease objective persuasiveness while increasing the individual's impression on recommender systems.

Keywords: Recommendation · Conversational · Persuasive

1 Introduction

In this study, we consider a recommender system that achieves persuasiveness by adopting conversational aspects. Persuasive technologies have been studied in various fields including item recommendation [3]. The task involves changing a user's conceptualization of a product or service. One method to change users' minds is to present an explanation for a recommendation instead of simply a list of offering choices. The assumption is that if the explanation is reasonable, then users can apply logic regarding an offer, and hence, the user may accept the offer. However, according to our observation, users are often reluctant to change their mind even though the offerings are logical. That is, in addition to the message being delivered, how they are delivered should be taken into account in designing recommender systems. When the recommender system is conversational, one approach to modify the message delivery style is to add characters to the system. In this paper, we examined if there are either positive or negative effects of added personal communication traits to conversational recommender systems.

Y.-H. Tseng et al. (Eds.): AIRS 2018, LNCS 11292, pp. 126–132, 2018.
https://doi.org/10.1007/978-3-030-03520-4_12

2   Related Work

Addition of characteristics to dialogue systems have been considered mostly by changing parameters reflecting predetermined generic personality dimensions such as degree of extraversion [7]. The change of behavior is considered to be a factor influencing personalities. For example, the effects of self-disclosure has been studies in a spoken dialogue system [9]. In this study, we considered text-based conversational systems and extract the specific profile of individuals.

3   Method

3.1   Reactive Token Generation (TokenGen)

To add personal communication traits, we employed two approaches. They are either non-lexical or lexical. For a non-lexical approach, we used the reactive token-based method [5]. This method extracts usage patterns of reactive tokens for a particular speaker in a corpus. Then, a probabilistic model for reactive token generation is created, which adds reactive tokens to the conversational system outputs. For example, tokens "I see" or "Uh-ha" are added. We call this method **TokenGen.** This method changes the manner in which information is delivered but does not change what is expressed.

3.2   Lexical Modification (LexMod)

For lexical modification, we adopted a machine translation-based method [8]. This method estimates probabilities of translation between a default word to the word that is peculiar to the speaker based on corpus statistics. Then, the utterances from the conversational system are modified based on the model. For example, "a nasty kid" may be changed into "a rude boy" and "Do you know?" may be changed into "You know, don't you?". Such replacement were conducted for function words in Japanese. We call this method **LexMod.** This method changes the manner in which information is expressed in a message. There were some unnatural utterances after automatic rewriting but we did not modified them manually.

3.3   Corpus

We need a corpus of different speakers as the basis for preparing personal communication traits. For the purpose, we used the Nagoya University Conversation Corpus (NUCC) [4] that is a transcribed corpus of Japanese natural conversation. The corpus contained the speakers' attributes, including their gender, age, and hometown. For the purpose of the pilot study, we extracted personal traits from three typical speakers: F1, F2, and M1. We selected both female (F) and male (M) speakers to compare difference in gender. Since there are more female speakers than male speakers (118 females and 20 males), we selected young and old female speakers (F1 and F2) for the purpose of examining age factors.

4   Experiment

4.1   Task

The task that the user is asked to carry out is restaurant ranking. First, task participants are asked to input preference information when selecting restaurants. The information consists of preferred food genre and preferred order of the restaurants' characteristics for selecting the relevant restaurant. For example, a user may describe the preference information as {(Genre), (Preference)} = {(Chinese), (food quality > service level > interior quality > price)} by using natural language. Second, the participants select five of thirty restaurants based on the information shown in the system interface restaurants' information panel (Fig. 1). Each restaurant is described based on its food quality, service level, interior quality, price, and genre. Then, the participants discuss their selection with the conversational recommender system that suggests re-ordering. This restaurant selection task is based on Andrews's work [2]. The recommendation can be conducted deterministically, not probabilistically. When there is logical inconsistency in user preferences based on the initial list of selected restaurants in terms of rankings, the system suggests changes to modify the ranking so that it conforms to the user-supplied preference data. For example, if a user stated that food quality was more important that the price range but the initial list was created based on the price range, the system asks to change the order of restaurants according to the food quality scores.

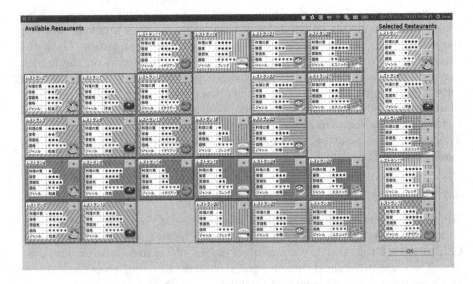

Fig. 1. Restaurant information provided to the task participants.

4.2 System Interface

The interface shown to the experiment participants is shown in Fig. 2. The interface is a text-based one; users type in the text to send a message (1) and receive feedback from the system through textual modality. Simultaneously, there is a visual display of the current restaurant ranking (6). The candidate restaurants and any selected restaurants are shown as tile displays in separate windows. The goal of the persuasive recommender system is to change the ranking so that it does not conform to the initially described participants preferences. To the system suggestions, user can react by clicking either the acceptance button or the rejection button (4). There is a confirmation button for their decisions (5). When users accept the suggestions by the system, they can change the ranking by clicking the upward arrow button or the downward arrow button (7). While users and the system converse, there are back-channeling utterances appear in the system utterance window (3).

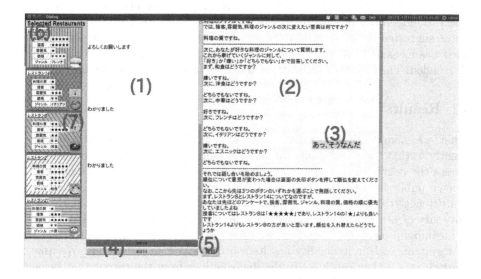

Fig. 2. Interface of the restaurant ranking system.

4.3 Experimental Procedure

The experiment was conducted as follows.

1. The task is explained using the dialogue system interface (on screen).
2. The participant select 5 favorite restaurants and rank them.
3. The participant answers the questionnaire on the importance of restaurant selection criteria.
4. The participant answers the questionnaire on the preferences for the cuisines.

5. The participant conducts interactions with the system. During the interaction session, the participant can change the restaurant ranking.
6. The participants answers the questionnaire on the system impression.

Each participant interact with one of three systems (TokenGen Method, LexMod Method, and Both Methods). Each participant evaluate both baseline default system (no individuality) and three different personal traits (F1, F2, and M1) extracted from the corpus. In the experiment, we have 26 participants. Among them, nine interacted with the system of TokenGen method, nine used the system of LexMod method, and eight experienced the system of both methods.

4.4   Evaluation Measure

Three measures were used for assessing the influence of added personal communication traits. The first is the objective persuasiveness, the second is subjective persuasiveness, and the third is the degree of satisfaction on the interactive session with the system. The objective degree of persuasiveness is measured by the ratio given by the occurrences of swapping the ranking by the users with those suggested by the system in the session. Further, the subjective persuasiveness was measured using a five-point scale questionnaire on the feeling of being persuaded. The degree of satisfaction was measure using questionnaire.

5   Results

First, we examined if our methods can add personal communication traits sufficiently so that the participants can feel the individualities from the systems. To assess the degree of personal trait representations when speakers are not well-known public figures and whose identities are not known in advance, utterance consistency had been used [6]. However, in our recommendation scenario, the diversity of utterances are limited and the consistency is not considered to be a meaningful measure. Therefore, we used the person identification test [5]. In our system, each individuality has its source speaker in the corpus. We asked participants if they can identify the dialogue logs from the corpus that belong to the source individuals after interacting with the system. The results of this experiment are shown in Table 1. Although the accuracy scores for F2 was not satisfactory, we assumed that they exceed the chance rates and added modification gives information on the source speakers.

The results of the experiments are shown in Table 2. We compared four settings. The first setting, baseline, use the default system output as it is. The second setting employed the TokenGen method, the addition of reactive tokens to the default system outputs. The third setting utilized the LexMod method, rewriting default output text by using machine translation. The fourth is the combination of both methods. Since the first method is non-lexical and the second method is lexical, they can be used simultaneously. As shown in the left column of Table 2, the baseline method achieved higher objective persuasiveness. Subjective persuasiveness and the degree of satisfaction are higher in the

Table 1. Estimation accuracies of individuals who are the source of added communication traits.

	F1	F2	M1
TokenGen method	0.89	0.22	0.56
LexMod method	0.67	0.56	0.78
Both methods	0.63	0.00	0.63

Table 2. Influences of added individualities in terms of re-ranking occurrences (objective) and system impression scores evaluated by the participants (subjective).

	Objective persuasiveness (%)	Subjective persuasiveness	Degree of satisfaction
Baseline	37	3.50	3.73
TokenGen method	36	3.56	4.07
LexMod method	35	3.26	3.63
Both methods	33	2.96	3.88

TokenGen method than the baseline and LexMod systems based on the five-point scale, as shown in the middle and right columns. It may be natural that adding the personalities that are not necessarily persuasive did not improve the degree of persuasiveness of the recommender systems. An interesting result is that even though the objective persuasiveness for TokenGen system is lower than the default system, the subjective persuasiveness is higher for the TokenGen system. We should examine the cause of this discrepancy further.

6   Conclusion

In this study, we compared two methods to introduce individuality in a restaurant recommender system in order to evaluate their persuasiveness. Individuality is realized by extracting personal communication traits from the face-to-face dialogue corpus of diverse speakers. The first method uses reactive token selection and the second method involves text rewriting. Experimental results on system users suggest that both methods do not improve objective persuasiveness but they do differ in subjective persuasiveness and degree of satisfaction.

There are several topics to be considered. The lexical method we employed was based on the utterance selections. There are attempts to modify utterance style after generating utterances [1]. It would be interesting to examine the influences of base dialogue systems. Also, for open domain dialogue system, various evaluation measures were considered [10]. Utilities of these metrics in our restricted recommendation scenario can be considered.

References

1. Akama, R., Inada, K., Inoue, N., Kobayashi, S., Inui, K.: Generating stylistically consistent dialog responses with transfer learning. In: Proceedings of the Eighth International Joint Conference on Natural Language Processing, pp. 408–412, Taipei, Taiwan (2017)
2. Andrews, P.Y.: System personality and persuasion in human-computer dialogue. ACM Trans. Interact. Intell. Syst. **2**(2), 1–27 (2012)
3. Fogg, B.J.: Persuasive Technology: Using Computers to Change What We Think and Do. Science & Technology Books (2002)
4. Fujimura, I., Chiba, S., Ohso, M.: Lexical and grammatical features of spoken and written Japanese in contrast: exploring a lexical profiling approach to comparing spoken and written corpora. In: Proceedings of the VIIth GSCP International Conference on Speech and Corpora, pp. 393–398 (2012)
5. Inoue, M., Ueno, H.: Dialogue system characterization by back-channelling patterns extracted from dialogue corpus. In: LREC, pp. 2736–2740 (2016)
6. Li, J., Galley, M., Brockett, C., Spithourakis, G., Gao, J., Dolan, B.: A persona-based neural conversation model. In: Proceedings of the 54th Annual Meeting of the Association for Computational Linguistics, vol. 1, pp. 994–1003 (2016)
7. Mairesse, F., Walker, M.: Personage: personality generation for dialogue. In: Proceedings of the 45th Annual Meeting of the Association of Computational Linguistics, pp. 496–503 (2007)
8. Mizukami, M., Neubig, G., Sakti, S., Toda, T., Nakamura, S.: Linguistic individuality transformation for spoken language. In: Lee, G.G., Kim, H.K., Jeong, M., Kim, J.-H. (eds.) Natural Language Dialog Systems and Intelligent Assistants, pp. 129–143. Springer, Cham (2015). https://doi.org/10.1007/978-3-319-19291-8_13
9. Ogawa, Y., Miyazawa, K., Kikuchi, H.: Assigning a personality to a spoken dialogue agent through self-disclosure of behavior. In: Proceedings of the Second International Conference on Human-agent Interaction. HAI 2014, pp. 331–337 (2014)
10. Venkatesh, A., et al.: On evaluating and comparing conversational agents. In: NIPS 2017 Conversational AI workshop, Long Beach (2018)

Guiding Approximate Text Classification Rules via Context Information

Wai Chung Wong[1(✉)], Sunny Lai[2,3], Wai Lam[1], and Kwong Sak Leung[2,3]

[1] Department of Systems Engineering and Engineering Management,
The Chinese University of Hong Kong, Shatin, N.T., Hong Kong
{wcwong,wlam}@se.cuhk.edu.hk
[2] Department of Computer Science and Engineering,
The Chinese University of Hong Kong, Shatin, N.T., Hong Kong
{slai,ksleung}@cse.cuhk.edu.hk
[3] Institute of Future Cities, The Chinese University of Hong Kong,
Shatin, N.T., Hong Kong

Abstract. Human experts can often easily write a set of approximate
rules based on their domain knowledge for supporting automatic text
classification. While such approximate rules are able to conduct classification at a general level, they are not effective for handling diverse and
specific situations for a particular category. Given a set of approximate
rules and a moderate amount of labeled data, existing incremental text
classification learning models can be employed for tackling this problem
by continuous rule refinement. However, these models lack the consideration of context information, which inherently exists in data. We propose a
framework comprising rule embeddings and context embeddings derived
from data to enhance the adaptability of approximate rules via considering the context information. We conduct extensive experiments and the
results demonstrate that our proposed framework performs better than
existing models in some benchmarking datasets, indicating that learning the context of rules is constructive for improving text classification
performance.

Keywords: Rule embedding · Context embedding · Text classification

1 Introduction

A common method for constructing text classification rules is to automatically
conduct supervised rule learning from labeled data. However, when the amount
of labeled data is of moderate size, pure supervised learning may not be effective
as expected. Besides labeled data, domain experts very often are able to write
a form of approximate classification rules based on their own knowledge easily.
These text classification rules typically detect word existence and nonexistence
in a document for determining the class label.

The work described in this paper is substantially supported by a grant from the Direct
Grant of the Faculty of Engineering, CUHK (Project Code: 4055093).

© Springer Nature Switzerland AG 2018
Y.-H. Tseng et al. (Eds.): AIRS 2018, LNCS 11292, pp. 133–139, 2018.
https://doi.org/10.1007/978-3-030-03520-4_13

While such approximate rules are able to conduct classification to some extent, one difficulty is that it is usually hard to blend conflicting rules into a unified model to handle various situations. One observation is that as words are combined to compose documents and paragraphs in different context, conflicts arise when it is used in dissimilar context. For instance, we may have the rule "IF the word 'goal' exists, THEN the document belongs to the 'sports' category.". However, it is not applicable to the document with the main content "The government's goal is to promote the tech industry to boost its economy." as this document should belong to the 'politics & government' category. If the context of the document is considered, one can automatically avoid the firing of such inaccurate rules for this particular document.

Given a set of approximate text classification rules and some labeled data, existing incremental text classification learning models can be employed for tackling this problem by learning the confidence score of the rules via exploiting the available approximate rules and the labeled data. For example, the Learn++ model proposed in [8] combines the final decision of weak classifiers by weighted majority voting. The method presented in [6] refines the model to handle the outvoting problem which arises when a new class is introduced. The model in [4] was later developed to handle possible changing environment. One common limitation of all the above methods is that they adjust the weighting of previously trained classifiers by their general performance, but without considering the use of context derived from data. This implies that the confidence scores of the rules are insensitive to the context information inherently found in data. We propose our framework which is capable of guiding approximate text classification rules to adapt to different situations according to their associated rule embedding and context embeddings derived from labeled data. Our model can also be regarded as rule adaptation or refinement based on context. We have conducted extensive experiments and the results show that our framework is better than existing methods in some benchmarking datasets.

2   Problem Definition

Assume there are some text documents and let x_i be a document represented by the word sequence $(w_1, w_2, ..., w_t)$ where $w_i \in V$ and V denotes the vocabulary set. We aim at determining the class label y_i for the corresponding document x_i where $y_i \in \mathcal{Y}$ and $\mathcal{Y}$ denotes the set of pre-defined class labels. We are also given a set of human provided classification rules $r_j \in \mathcal{R}$, $j = 1, ..., M$. Each r_j takes the form $r_j : c_j \Rightarrow l_j$, where c_j is the *condition* and l_j is the class label. c_j is represented by a set of words $\{w_1, ..., w_n\}$ denoting the existence of w_i as a condition. $\mathcal{R}$ can be regarded as a set of *approximate* text classification rules. Such form of rules are typically easily provided by human experts and these rules are able to conduct classification to some extent. In practice, the condition of each rule often consists of 1 to 2 words. In addition to $\mathcal{R}$, we are given a set of labeled dataset (x_k, y_k) where x_k is the kth document; y_k is the corresponding class label of x_k. Our goal is to distill the quality of approximate text classification rules by making use of labeled data of moderate size such that the model can achieve better classification performance than the original $\mathcal{R}$.

3   Our Proposed Model

3.1   Model Description

Figure 1 illustrates an overview of our model. We introduce rule embedding e_{r_j}, a vector representation of the rule r_j, and context embedding e_{x_k}, a vector representation of the document x_k. Armed with these embeddings, we can compute the confidence score based on them and rank the rules. Our model consists of two main components—Context Encoder and Rule Matcher. Context Encoder is used for extracting the context embeddings from the documents. Rule Matcher analyzes the context embedding and rule embeddings with the goal of making the prediction of the class label.

Context Encoder. As the context of a document is governed by its surrounding words, the ability of summarizing the word sequence of a document and transforming it into a compact and comparable context embedding is useful. We use a recurrent neural network model known as Long Short Term Memory (LSTM) [5] architecture as the core model of Context Encoder for encoding document-level context. We first map each word w_i of the document x_k to a word embedding e_{w_i}. Then the word embedding is passed them to a LSTM cell. Finally we retrieve the last hidden state of LSTM as the context embedding e_{x_k} for document x_k.

Rule Matcher. Recall that the Rule Matcher is to analyze the context embedding and rule embeddings, we define $r_j^* : c_j^* \Rightarrow l_j^*$ as a *matched rule* of the document x_k if x_k fulfills the condition c_j^*, i.e. c_j^* is a subset of x_k. Correspondingly, we denote $\mathcal{R}_k^*$ as the set of matched rules of x_k. We employ the dot product between the context embedding and the rule embedding as the confidence score of the associated rule. The Rule Matcher component first finds $\mathcal{R}_k^*$ from $\mathcal{R}$, and maps them to their corresponding embeddings, i.e. $e_{r_j^*}$ where $r_j^* \in \mathcal{R}_k^*$. With the context embedding e_{x_k} extracted from Context Encoder, we are able to compute the confidence score $d_{r_j^*}$ of r_j^* as follows:

$$d_{r_j^*} = e_{x_k}^T e_{r_j^*} \qquad (1)$$

After obtaining the confidence scores, we can rank each r_j^* according to their $d_{r_j^*}$ and output the class label l_j^* with the highest $d_{r_j^*}$ as the final prediction for x_k. For example, consider a document x_1 containing 3 words for simplicity, $(learn, play, tennis)$. Suppose that there are 3 rules, $r_1 : \{learn\} \Rightarrow Education$, $r_2 : \{stock\} \Rightarrow Finance$ and $r_3 : \{tennis, paly\} \Rightarrow Sports$. Rule Matcher first finds $\mathcal{R}_1^*$ which is $\{r_1^*, r_3^*\}$ because x_1 fulfills the condition of r_1 and r_3. Assume that the rules embeddings $e_{r_1^*}$ and $e_{r_3^*}$ have been learned in the training procedure for r_1 and r_3 respectively. Suppose that the context embedding e_{x_1} is obtained from Context Encoder. The Rule Matcher component then computes the confidence scores $d_{r_1^*}$ and $d_{r_3^*}$ by taking the dot product as shown in Eq. 1. Suppose that $d_{r_1^*}$ is larger than $d_{r_3^*}$. The Rule Matcher component outputs l_1^*, namely *Education*, as the prediction of x_1.

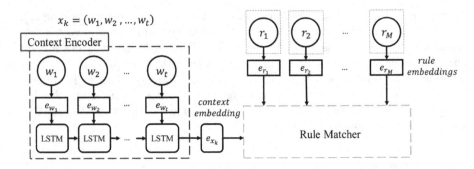

Fig. 1. Overview of our proposed model

3.2 Training

The purpose of training is to learn the model parameters of LSTM, word embeddings of each word and the rule embeddings of each rule. It can be observed that if r_j^* has the correct class label for the document x_k, it should have higher confidence score than those having wrong labels. Using the example in the previous section, if we also know that $y_1 = Sports$, $d_{r_3^*}$ should have a higher value than $d_{r_1^*}$ as r_3^* has the correct class label for x_1. Here, we first define $d_{r_j^*,l_j^*=y_k}$ and $d_{r_j^*,l_j^*\neq y_k}$ as the confidence scores of r_j^* which has the correct and wrong class label for the document x_k respectively. Our goal is to train our model to capture the relation of $d_{r_j^*,l_j^*=y_k}$ and $d_{r_j^*,l_j^*\neq y_k}$ which is $d_{r_j^*,l_j^*=y_k} > d_{r_j^*,l_j^*\neq y_k}$. In order to accomplish this goal, we separate $\mathcal{R}_i^*$ for each x_k into all possible $(d_{r_j^*,l_j^*=y_k}, d_{r_j^*,l_j^*\neq y_k})$ pairs, and minimize the margin ranking loss of all these pairs by enumeration until the loss converges. The margin ranking loss L is given by

$$L(d_{r_j^*,l_j^*=y_k}, d_{r_j^*,l_j^*\neq y_k}) = max(0, -d_{r_j^*,l_j^*=y_k} + d_{r_j^*,l_j^*\neq y_k} + \phi) \qquad (2)$$

where ϕ is a parameter for setting the minimum margin required between two confidence scores.

4 Experiments

4.1 Dataset

Our experiments were performed on two benchmarking datasets, namely Reuters and Yahoo! Answers. The Reuters dataset is obtained from UCI machine learning repository[1] and Yahoo! Answers dataset is obtained from [9]. For Reuters, we extract the five most abundant classes from the dataset. The total numbers of training and testing samples are 5787 and 2298. Note that the dataset is imbalanced. The most frequent class has 2848 training samples while the lowest has 359 training samples. For Yahoo! Answers, 10 largest main classes are extracted. We select a subset to simulate a practical scenario where the availability of moderate size of training data. We also prepared a balanced dataset from this corpus. The total numbers of training and testing samples are 9330 and 60000.

4.2   Experimental Setup

We invite a graduate student with good knowledge of Reuters and Yahoo! categories as a human expert to write approximate classification rules. For documents preprocessing, the documents in the datasets are processed by lemmatization using spaCy. For models implementation, the neural network related models are implemented using PyTorch [7]. We use the classification accuracy as our evaluation metric.

For comparison models, we adopt three representative models, namely, the approximate classification rules, incremental learning method Learn++, and LSTM. For using the approximate rules, we adopt a majority voting strategy. For example, if a document matches 5 rules where 2 rules are associated with class 1 and 3 rules are associated with class 2, the final prediction for this document would be class 2.

We implement Learn++. First, we treat the approximate classification rules as the initial classifier. Since Learn++ requires a supervised learning model as the BaseClassifier, we choose an automatic rule induction algorithm RIPPER [2] as the BaseClassifier. The reason of employing RIPPER is that it is a rule based model and has been shown to perform well in text classification tasks [3]. After that we follow the same procedure of the Learn++ algorithm.

For the LSTM model, we first pass the context embedding to the fully connected layer. Then we apply softmax function to obtain the class label distribution. Next, we minimize the cross entropy between the class label distribution and the expected distribution.

4.3   Experimental Results

Table 1 shows the text classification accuracy (in percentage) of two datasets in our experiments. We observe that the approximate classification rules are able to conduct classification to a certain degree. Learn++ achieves comparable results against the approximate rules by making use of the labeled data to generate new rules and adjust the weightings of two sets of rules. LSTM performs better than approximate rules as well as Learn++ as it is capable to consider context information from labeled data when making the prediction. Finally, our model

---

[1] https://archive.ics.uci.edu/ml/datasets/reuters-21578+text+categorization+ collection.

outperforms all these models because we utilize the approximate rules and context information. The results demonstrate that our model is able to improve the quality of the rules and achieve better performance.

Table 1. Experimental results. † and ◇ denote that it is statistically significant to conclude the improvement of our model over LSTM and Learn++ respectively.

Model	Reuters	Yahoo
Approx. rules	94.56	49.93
LSTM	96.30	50.03
Learn++	95.00	49.93
Our approach	**97.56**$^{†◇}$	**58.47**$^{†◇}$

To further investigate our model, we performed McNemar's Test [1], a statistical significance test for comparing classification models. We conduct McNemar's tests for two comparisons (1) our model against LSTM and (2) our model against Learn++. The results show that all the improvements of our model are statistically significant with 99% confidence level, indicating that our model is better than either LSTM or Learn++.

5   Conclusions

We propose a framework comprising rule embeddings and context embeddings derived from data to enhance the adaptability of approximate rules via considering the context information. We report experimental results on two benchmarking datasets. Our experiments indicate that by considering context information, rules are more adaptive in diverse situations. This encourages us to explore how advanced neural network methods can help improve rules with context in future research.

References

1. Bostanci, B., Bostanci, E.: An evaluation of classification algorithms using Mc Nemar's test. In: Bansal, J., Singh, P., Deep, K., Pant, M., Nagar, A. (eds.) BIC-TA 2012. AISC, vol. 201, pp. 15–26. Springer, India (2013). https://doi.org/10.1007/978-81-322-1038-2_2
2. Cohen, W.W.: Fast effective rule induction. In: Twelfth International Conference on Machine Learning (ICML), pp. 115–123 (1995)
3. Cohen, W.W., Singer, Y.: Context-sensitive learning methods for text categorization. ACM Trans. Inf. Syst. (TOIS) **17**(2), 141–173 (1999)
4. Elwell, R., Polikar, R.: Incremental learning of concept drift in nonstationary environments. IEEE Trans. Neural Netw. **22**(10), 1517–1531 (2011)
5. Hochreiter, S., Schmidhuber, J.: Long short-term memory. Neural Comput. **9**(8), 1735–1780 (1997)

6. Muhlbaier, M.D., Topalis, A., Polikar, R.: Learn++. NC: combining ensemble of classifiers with dynamically weighted consult-and-vote for efficient incremental learning of new classes. IEEE Trans. Neural Netw. **20**(1), 152–168 (2009)
7. Paszke, A., et al.: Automatic differentiation in PyTorch. In: Advances in Neural Information Processing Systems (NIPS) - Workshop (2017)
8. Polikar, R., Upda, L., Upda, S.S., Honavar, V.: Learn++: an incremental learning algorithm for supervised neural networks. IEEE Trans. Syst. Man Cybern. part C (Appl. Rev.) **31**(4), 497–508 (2001)
9. Zhang, X., Zhao, J., LeCun, Y.: Character-level convolutional networks for text classification. In: Advances in Neural Information Processing Systems (NIPS), pp. 649–657 (2015)

Medical and Multimedia

Key Terms Guided Expansion for Verbose Queries in Medical Domain

Yue Wang$^{(\boxtimes)}$ and Hui Fang

Department of Electrical and Computer Engineering, University of Delaware,
Newark, Delaware, USA
{wangyue,hfang}@udel.edu

Abstract. Due to the complex nature of medical concepts and information need, the queries tend to be verbose in medical domain. Verbose queries lead to sub-optimal performance since the current search engine promotes the results covering every query term, but not the truly important ones. Key term extraction has been studied to solve this problem, but another problem, i.e., vocabulary gap between query and documents, need to be discussed. Although various query expansion techniques have been well studied for the vocabulary gap problem, existing methods suffer different drawbacks such as inefficiency and expansion term mismatch. In this work, we propose to solve this problem by following the intuition that the surrounding contexts of the important terms in the original query should also be essential for retrieval. Specifically, we first identify the key terms from the verbose query and then locate the contexts of these key terms in the original document collection. The terms in the contexts are weighted and aggregated to select the expansion terms. We conduct experiments with five TREC data collections using the proposed methods. The results show that the improvement of the retrieval performance of proposed method is statistically significant comparing with the baseline methods.

1 Introduction

Biomedical records and medical database are valuable resources for both patients and physicians. However, due to the nature of the complexity of the biomedical domain, the queries are more complicated than the ones from web domain. For instances, physicians may issue the query *"patients with flu vaccines who take cough and fever medicine"* when they need to find those particular patients. The medical queries are complicated for two reasons. On the one hand, the concepts in medical domain are more complex by itself. Such as the"flu vaccines", more than one term are needed to describe one medical concept. On the other hand, usually more than one aspect are required in the query. As shown in the example query, the inclusion criteria contain two aspects, which makes it longer. The average query length for the medical tracks in TREC is more than 20 terms per query [1], which is much longer than the average web query, even longer than the verbose web queries [2].

© Springer Nature Switzerland AG 2018
Y.-H. Tseng et al. (Eds.): AIRS 2018, LNCS 11292, pp. 143–156, 2018.
https://doi.org/10.1007/978-3-030-03520-4_14

The verbose queries could be problematic within the framework of current search engine, since the search engine will promote the documents covering every query term, but not the ones which are truly important. Therefore, those noise terms will affect the overall retrieval performance of verbose queries. Key term identification from verbose queries has been studied by existing work [2–4]. Although different methods have been proposed to solve this method, the intuition is the same, i.e., extract the useful terms from the verbose queries to reformulate the original queries. However, even with the identified key terms, there is still another problem yet to be solved, i.e., how to bridge the vocabulary gap between the ones used in the query and in the documents.

Query expansion is well studied to solve the problem of vocabulary gap. Query expansion techniques could be divided as internal query expansion and external query expansion based on the resources used for expansion. Internal query expansion selects the expansion from the document collection itself, so the quality of the expansion term is guaranteed. However, since a preliminary retrieval task is needed to select the candidate expansion terms, the run time is very slow. External query expansion techniques rely on certain knowledge bases to select expansion term. For the terms occurred in the original query, related terms are chosen based on predefined rules, such as synonyms. Clearly the external expansion has a faster run time, but the correctness of selected term is questionable since it depends on the reliability of the selected resources and the term usage could be different. However, most of the existing query expansion methods geared towards the keyword queries, which can not be directly applied to the verbose queries.

In this work, we studied the problem of query expansion for verbose queries in the medical domain. To the best of our knowledge, we are one of the first few groups to focus on this problem. Intuitively, the contexts showing together with the key terms from the verbose queries should receive more weights than the ones show farther away from the key terms. For instance, given a long document which the query term only occurs in the first paragraph, the terms in the first paragraph should be more relevant to the original query than the ones from the end of the document. Under this intuition, we proposed the key terms guided query expansion method to solve the query expansion for verbose queries. Specifically, we first identify the key terms in the verbose query. Those key terms are then guide us to locate the contexts, and expansion terms will be selected and weighted from these contexts. Since the same data collection is used to select the expansion term as the retrieval task, the correctness of the selected expansion term is guaranteed. In addition, our method has a shorter run time than the internal query expansion since no preliminary run is required. We evaluated our methods using five data collections released by TREC. The results show that our method could improve the retrieval performance significantly.

2   Related Work

Our method is closely related to verbose query processing, query expansion, and medical domain retrieval. Numerous work have been done in these directions and we will introduce the closet ones in this section.

Verbose Query Processing. Selecting important terms from verbose query is not a novel task in information retrieval domain. Turney first introduced the concept of useful information extraction from verbose query back in 2000 [5]. Since that, researchers explored different aspects of the verbose query processing, such as query reductions [4,6,7], query term weighting [8–10], query reformulation [11,12], etc. However, most of the existing works focus on web search. The proposed method may not work on medical domain queries since the query length are very different. Our previous study proposed a machine learning based key term identification method for verbose medical queries [3]. In particular, we proposed 16 features that used to select useful terms from verbose query. In this paper, we focuses on how to utilize the identified key terms for query expansion.

Query Expansion. Query expansion techniques aim to improve retrieval performance by reformulating the original query. Query expansion methods are either relies analysis of the document collection itself [13], or external knowledge bases [14]. For internal query expansion, both relevance feedback and pseudo relevance feedback methods have been proven to be effective [15–17]. However, pseudo relevance feeback methods cannot improve the retrieval perforamnce for every query since the top ranked documents might not be relevant [13,18]. On the other hand, the effectiveness of external query expansion highly relies on the robustness the selected resources [19]. In addition, the different term usage of the external resources could also lead to non-optimal expansion terms. There are several work studied proximity information for query expansion. Xu and Croft used a feature based selection method to select expansion terms with in a document passages [20]. However, the expansion terms are considered as binary decision in their work. We argue that this binary representation is not the best model. Lv and Zhai proposed the positional relevance model which incorporates the proximity information with the probabilistic language model for feedback [21]. Ermakova et al. studied a PRF based method by incorporating distance information into LM formalism [22]. They capture the distance in terms of sentences but not term. Our work is different from these two studies in two aspects. First, we measure the distance in terms of raw terms. Second, we studied the problem in the medical domain while they focused on web domain. More recently, Zamani and Croft proposed a machine learning algorithm to learn the word embedding based on the notion of relevance [23]. The experiment results show that the terms selected by their model is more relevant to the whole query than the baseline models. However, their model is trained using the general web query logs which makes it inapplicable for this domain specific task.

Medical Domain Retrieval. Information retrieval in bio-medical domain has been a trending topic with the growth of the health-care market. Based on how documents and queries are represented, existing work could be divided into term based representation and concept based representation. Most of the work in the term based representation are built on the bag-of-word assumption and different techniques are applied to improve the performance [24,25]. Concept based representation, on the other hand, treat the documents and queries as bag-of-concepts. Domain specific NLP toolkits, such as MetaMap and cTAKEs, are used to map the raw text to medical concepts. Some researchers then apply the term based retrieval models to the concept based collections to evaluate the performance [26,27], while others studied the possible weighting strategy for concept based representation [28]. However, few work in this direction focused on dealing with the verbose queries as our work.

3   Methods

The fundamental goal of query expansion is to reformulate the query in order to improve the retrieval performance. Including more relevant terms to the original query is one of the commonly used methods. However, the most obvious difficulty for query expansion on verbose query is that the less important terms would result in drifting of the expansion techniques. In another word, the terms that are only related to the less important query terms would dominate the original query. To solve this problem, we propose to first identify the important terms from the verbose query, and apply query expansion techniques only on those key terms. In terms of the query expansion method, in order to overcome the drawbacks of the existing methods, we propose to utilize the contexts in the document collection of the identified key terms as the pool to select the expansion terms. The intuition is that the terms occur closer to the key terms should be favoured more than ones show remotely.

The proposed system can be illustrated as shown in Fig. 1. The original query requires retrieving the documents related to *"patients with flu vaccines who take cough and fever medicine"*. Clearly, the important terms, i.e., key terms, in this query are *"flu vaccines"* and *"cough and fever medicine"*. Note that the important terms could be both a single term, such as *"PTSD"*, or a phrase, such as *"flu vaccines"*. For the sake of simplicity, we use key term to refer both cases. These two key terms, which represent the core meaning of the original query, are identified in the first step and will be used in the following steps. We applied the method discussed in [3] for this step. Specifically, we trained a classifier using logistic regression with the features introduced in their paper. We applied this model to the verbose queries to extract the important terms. For each key term, we then locate all the contexts in the original document collections. Those contexts, as shown on the plot, are used to generate candidate expansion terms. The proximity information is considered and the weight of the expansion terms is assigned based on that in the last step.

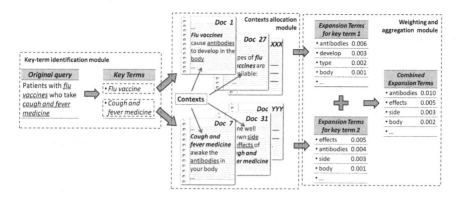

Fig. 1. Flowchart of the proposed system

3.1 Expansion Term Weighting and Aggregation

Carefully selecting the expansion terms is the key to the success of our proposed system, which requires a convincing weighting schema for each candidate term. In this section, we discuss three weighting schema for expansion term weighting.

Formally, we would like to define some notations first. For a given verbose query Q, there are N identified key terms (KT). For each key term KT_i, we located M_{KT_i} surrounding contexts (S_{KT_i}) in total. For each surrounding contexts $S_{KT_i}^j$, we define its length as k. We further define the function $dis_{S_{KT_i}^j}(t_{exp}, KT_i)$ that captures the distance from the target candidate expansion term to the key term in the jth contexts of key term KT_i. The goal is to find the optimal weight $Weight(t_{exp})$ for each candidate expansion term t_{exp}.

The most straight-forward weighting schema would be treating each candidate term equally. We denote this method as **Uniform**. Formally, the weight of t_{exp} could be defined as follows:

$$Weight_U(t_{exp}) = \sum_{KT=1}^{N} \frac{1}{M_{KT_i}} \sum_{S_{KT_i}^j=1}^{M_{KT_i}} \frac{Occur(t_{exp}, S_{KT_i}^j)}{k} \qquad (1)$$

where the $Occur(t_{exp}, S_{KT_i}^j)$ is the count of the term t_{exp} shows in the jth surrounding texts of key term KT_i.

One limitation of the Uniform method is that the proximity information of the candidate expansion term is not considered in the weighting schema. This is counter-intuitive since that, ideally, the candidate expansion term occurs closer to the key terms should be more important than the ones show remotely. In order to capture this intuition, we propose the following two new weighting schemata:

Normal: One possible approach of the importance of the candidate expansion terms with respect to the distance would be the standard normal distribution with the position of the key term as its mean. Formally, the weight of each candidate expansion term could be described as:

$$Weight_N(t_{exp}) = \sum_{KT=1}^{N} \frac{1}{M_{KT_i}} \sum_{S_{KT_i}^j=1}^{M_{KT_i}} \frac{1}{\sqrt{2\pi}} e^{-\frac{\left(\frac{dis_{S_{KT_i}^j}(t_{exp},KT_i)*3}{k}\right)^2}{2}} \tag{2}$$

The constant 3 in the exponent guarantees that the distance factor is in the 99% confidence range of the normal distribution. Therefore we could fully use the range to weight the candidate expansion terms.

Reciprocal: The problem of the normal distribution is that it the weight different for the terms closer to the key terms are not significant. Therefore, we also proposed the reciprocal model to capture the importance of expansion term with the locality information. Specifically:

$$Weight_R(t_{exp}) = \sum_{KT=1}^{N} \frac{1}{M_{KT_i}} \sum_{S_{KT_i}^j=1}^{M_{KT_i}} \frac{dis_{S_{KT_i}^j}(t_{exp},KT_i)}{k} \tag{3}$$

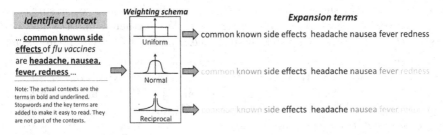

Fig. 2. Effectiveness of the weighting schema. The darker the expansion term is, the more weight it is assigned.

The three weighting schema differs from each other on how to utilize the locality information of the candidate terms to the key terms. These three weighting schema are illustrated in Fig. 2. As can be seen, the reciprocal method penalize the remote terms the most, which follows the expansion intuition very well. We hypothesis that the Reciprocal and Normal weighting schema will outperform the Uniform method since the proximity information should be considered.

3.2   Contexts Allocation

The success of expansion term selection not only relies on candidate term weighting, but also the quality of the contexts allocation. The mis-identified contexts could introduce more noise terms than useful terms. Therefore, contexts allocation is also a critical step. Due to the unstructured nature of the natural language, simply match the key terms as it shown in the query would lead to a sub-optimal performance since a lot of occurrence would be lost. By observing how the key terms are used in the documents, we noticed that people tend to switch the orders of the words in the key terms or add other terms to the key terms. Thus, we studied the effects of the gap size and order preserve when locating the surrounding texts.

Order Among Terms. It is natural to assume that we should reserve the term order when we locate them in the original documents. This is true in some cases since the syntactical order among terms could carry useful information. However, for the other cases, such as "cough and fever medicine" and "fever and cough medicine", the order among terms is not a critical requirement as the others. Therefore, it is worth validating how the terms order affects the text window allocations.

Gap Size. We refer the extra terms occur between the closet two identified key terms as gap terms. If the phrase in the documents matches the key terms exactly, the gap of this match is 0. For example, if the key term is "flu vaccines", the phrase "flu viruses vaccines" is a match of the original key term with gap size 1. Obviously, the smaller gap size will generated a more accurate candidate term list.

Context Length. The previous two parameters control how to locate the key term in the original documents, while the length of the context is also a critical parameter to consider. Intuitively, the longer the context is, the more possible promising expansion term we could identify. However, it is also true that, with the increase of the context, more noise terms would pollute the pool of candidate expansion terms. Therefore, a balanced context length need to be decided for the optimized retrieval performance. In our experiment, we define the $k/2$ as the number of total terms on either side of the occurrence of the key terms in the original document. Therefore, the total number of candidate expansion terms included in each occurrence will be k. In addition, the stopwords are skipped when we select the context.

4 Experiments

4.1 Experiment Setup

We utilized the data collections from TREC Medical Records track and Clinical Decision Support track to test the effectiveness of our proposed methods. The Medical Record track (MED) was organized in 2011 and 2012. There are 85 queries released in total for these two years. These queries are directly used in our experiments. 30 queries are released for each year's for Clinical Decision Support (CDS) track from 2014 to 2016. The average query length for each data collection is reported in Table 1. As it can be seen, even the collection with the shortest average query length, i.e, MED12, the queries are still longer than the verbose query in the web domain.

Table 1. Average query length (number of terms) for different collections.

Data sets	MED11	MED12	CDS14	CDS15	CDS16
Avg. query length	9.82	9.06	26.97	21.87	34.4
# of documents	17,198	17,198	732,980	732,980	1,255,260
Avg. document length	2659	2659	3488.4	3488.4	5619.6

We used Dirichlet prior smoothing [29] as the baseline retrieval function. Although we evaluated our method with Dirichlet retrieval function, we want to point out that our method can be applied to any other retrieval functions. For the baseline method, we used the original query as it is, without modification and query expansion. We denote this method as **Baseline**. In addition, we also included two state of the art methods for query expansion. Pseudo Relevance Feedback (PRF) is known to be an effective query expansion technique. We used the language model based method introduced by Lavrenko and Croft [30], which use the top ranked documents to compute the relevance model and then to select the expansion terms. This method is denoted as **PRF**. Word2vec [31], which is word embedding models published by Google, is also a strong baseline. We trained the word2vec model using the bio-medical data collections so the generated word embedding would be concentrated in the domain. For each identified key terms, we retrieve the top 20 terms in the term vector. The weights from different key terms in the query are combined to generate the expansion terms. We denoted this methods as **W2V-bio**. For both the two baseline methods and our methods, we selected the top 20 expansion terms for each query and combined them with the original query with the weight on original query set to 0.7. These parameters are set based on the best performance tuned on the CDS14 data set. The weights on the expansion terms are also kept to regularize the expansion terms. The performance of the MED11 data collection is reported based on MAP, while infNDCG is reported for the other data collections, which are the official measures used in the corresponding years.

4.2 Query Expansion Results

We discuss the performance of the proposed system in this section. We first report the optimal performance with best tuned parameters and then discuss the robustness of the system by reporting the train-test performance.

With the best tuned parameters, the performance of each method is reported in Table 2. We could easily see that our methods outperform the baseline method that directly utilizing the original query by comparing the performance in Table 2. The improvement is statistically significant. The Reciprocal methods

Table 2. Retrieval performance using different methods. The [†] and [‡] indicate the improvement over *PRF* and *W2V-bio* is statistically significant at 0.05 level based on Wilcoxon signed-rank test, respectively.

	MED11	MED12	CDS14	CDS15	CDS16
Baseline (No FB)	0.3474	0.4695	0.1656	0.1969	0.1792
PRF	0.3579	0.4785	0.2045	0.2437	0.2343
W2V-bio	0.3628	0.4849	0.1912	0.2361	0.1894
Uniform	0.3612	0.4713	0.2197	0.2375	0.2501
Normal	0.3606	0.4764	0.2281	0.2505	**0.2613**[†,‡]
Reciprocal	**0.3808**	**0.5081**[†]	**0.2362**[†,‡]	**0.2554**	0.2567[†,‡]

is the best method among the three proposed weighting schema, which indicates that the distance to the key term is a strong signal when selecting the expansion terms. The closer the candidate term located to the key term, the higher the weight should be. In addition, we could also see that the Reciprocal and Normal weighting schema significantly improve the performance on some of the date collections comparing with the two strong baseline expansion techniques, which shows that weighting the terms based on locality is better.

Table 3. Testing performance using parameters trained on the other four data collection. The [†] and [‡] indicate the improvement over *PRF* and *W2V-bio* is statistically significant at 0.05 level based on Wilcoxon signed-rank test, respectively.

	MED11	MED12	CDS14	CDS15	CDS16
Baseline (No FB)	0.3422	0.4572	0.1592	0.1901	0.1685
PRF	0.3511	0.4612	0.1989	0.2414	0.2267
W2V-bio	0.3523	0.4745	0.1838	0.2282	0.1804
Uniform	0.3583	0.4591	0.2134	0.2284	0.2435
Normal	0.3496	0.4681	0.2172	0.2380	**0.2562**[†,‡]
Reciprocal	**0.3718**	**0.4923**[†]	**0.2275**[‡]	**0.2498**	0.2471[†,‡]

We also conduct cross validation to test the robustness of the parameters. Specifically, we train the parameters on four data collection and test the performance on the fifth one. We rotated the testing collection five times, so each collection will be used as test collection once. The results are reported as Table 3. Clearly, the reciprocal weighting schema is still the most effective weighting schema. In addition, the testing performance is very close to the best tuned performance, which indicates the proposed are robust with respect to the parameter setting.

4.3   Parameter Sensitivity

As discussed in the previous section, the performance of our methods depends
on several parameters. We will discuss the parameter sensitivities in this section.
We report the performance as shown in Fig. 3. Although only the performance on
CDS collections are reported in the plot, we want to point that the performance
on Med collections are consistent with CDS collections.

Gap Size and Order Among Terms. As can be seen from Fig. 3(a), the
performance drops significantly with the increase of the gap size. This is expected
since when the gap size increases, the term located in the original document may
not be the actual query key terms. For instance, for the key term *"quit drinking"*,
the phrase *"... patient quits smoking but starts drinking ..."* will be identified
as a candidate phrase when gap window is set to 3, although we could see that
the phrase is clearly not relevant to the original key terms. The order among
the terms has limited effect on the retrieval performance. We believe this is
because the occurrence of order reversing is not common enough to affect the
performance.

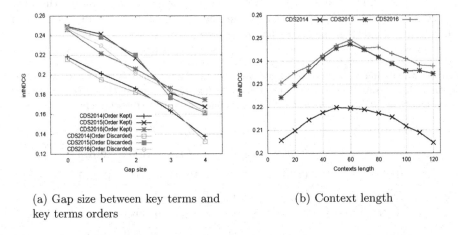

(a) Gap size between key terms and
key terms orders

(b) Context length

Fig. 3. Retrieval performance with different parameters.

Context Length. We report the performance changes with respect to the text
window size as shown in Fig. 3(b). We could see that with the increase of the
window size, the performance reach the peak when the window size is around 50
to 60 terms. With a further investigate on the expansion term, we noticed that
when the context length is small (i.e., 10 to 30 terms), the expansion terms are
mostly syntactic, but not semantic, related to the key terms. On the other hand,
when the context length becomes larger, more noise terms will be included in
the expansion.

4.4   Further Analysis

We argue the proposed methods are designed to solve the problem of query expansion for verbose queries. Therefore, we break down the overall performance into different query length to check how our methods perform. The Fig. 4 shows the improvement of our method (Reciprocal) and the two query expansion methods comparing with the baseline (without feedback) methods.

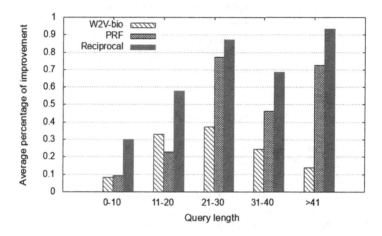

Fig. 4. Average improvement of proposed method comparing with baseline methods with different query length

As can be seen from Fig. 4, the improvement of our method over the baseline is clear and consistent, especially with the more verbose queries (i.e., longer than 10 terms). By comparing with two state-of-the-art query expansion methods, we can see that our method outperform the Google word2vec method significantly on the longest group. The improvement over the PRF method is also clear on every query length group, which indicates that the expansion terms selected by our method are more effective than the other two methods.

We further studied the expansion terms selected by different methods. The results are shown in Table 4. Only the top 10 expansion terms together with the weight for each term is shown. The PRF and W2V-bio methods improve the performance to 0.1376 and 0.1268, respectively, while the proposed reciprocal method boosted the performance to 0.2674. By comparing the selected expansion terms and the weights for each term for different method, we could find out that our method could not only bring more relevant term, such as anoxia, but also can weight those expansion term correctly.

Table 4. Expansion terms comparision among different methods for original query "*30-year-old woman who is 3 weeks post-partum, presents with shortness of breath, tachypnea, and hypoxia*" (The useful terms are highlighted)

Query type	Query content
PRF	Air, patients, rib, heart, **hypotension,** report, count, chest, birth, surgery
W2V-bio	**Breathless,** nose, exercise, run, **oxygen,** affect, **hypotension,** patients, **ischemia, lung**
Reciprocal	**Lung, hypotension,** heart, **oxygen, anoxia, ischemia,** study, shallow, blood, chest

5   Conclusions and Future Work

In this work, we proposed a novel query expansion method for verbose query in medical domain. To the best of our knowledge, we are one of the first few groups studied this problem. Specifically, surrounding texts for the key terms identified in the verbose query are used to generate the candidate expansion terms. The candidate expansion terms are then weighted based on the proximity information with respect to the corresponding key terms. Experimental results over five data collections show that using the selected expansion terms are more effective than the ones selected by Google word2vec and PRF, and our method could significantly improve the retrieval performance over the baseline methods, especially for longer queries.

We are interested in several directions as future work. First, we would like to explore more weighting schema. Second, evaluating the effects of the key term identification step would be an interesting topic to study. Other verbose query processing technique could be applied to select the key terms in the first sub-module. Finally, it would be interesting to study how to leverage the relations among all the key terms to regularize the weighting of expanded query terms.

Acknowledgements. This research was supported by the U.S. National Science Foundation under IIS-1423002.

References

1. Roberts, K., Demner-Fushman, D., Voorhees, E.M., Hersh, W.R.: Overview of the TREC 2016 clinical decision support track. In: TREC (2016)
2. Gupta, M., Bendersky, M.: Information retrieval with verbose queries. In: Proceedings of the 38th International ACM SIGIR Conference on Research and Development in Information Retrieval. SIGIR 2015, pp. 1121–1124 (2015)
3. Wang, Y., Lu, K., Fang, H.: Learning2extract for medical domain retrieval. In: Sung, W.K., et al. (eds.) Information Retrieval Technology. LNCS, vol. 10648, pp. 45–57. Springer, Cham (2017). https://doi.org/10.1007/978-3-319-70145-5_4

4. Bendersky, M., Croft, W.B.: Discovering key concepts in verbose queries. In: Proceedings of the 31st Annual International ACM SIGIR Conference on Research and Development in Information Retrieval, SIGIR 2008, pp. 491–498 (2008)
5. Turney, P.D.: Learning algorithms for keyphrase extraction. Inf. Retrieval **2**(4), 303–336 (2000)
6. Balasubramanian, N., Kumaran, G., Carvalho, V.R.: Exploring reductions for long web queries. In: Proceedings of the 33rd International ACM SIGIR Conference on Research and Development in Information Retrieval. SIGIR 2010, pp. 571–578 (2010)
7. Wang, Y., Fang, H.: Extracting useful information from clinical notes. In: TREC 2016 (2016)
8. Paik, J.H., Oard, D.W.: A fixed-point method for weighting terms in verbose informational queries. In: Proceedings of the 23rd ACM International Conference on Conference on Information and Knowledge Management, pp. 131–140 (2014)
9. Zhao, L., Callan, J.: Term necessity prediction. In: Proceedings of the 19th ACM International Conference on Information and Knowledge Management, CIKM 2010, pp. 259–268 (2010)
10. Lease, M.: An improved Markov random field model for supporting verbose queries. In: Proceedings of the 32nd International ACM SIGIR Conference on Research and Development in Information Retrieval, SIGIR 2009, pp. 476–483 (2009)
11. Bonchi, F., Perego, R., Silvestri, F., Vahabi, H., Venturini, R.: Recommendations for the long tail by term-query graph. In: Proceedings of the 20th International Conference Companion on World Wide Web, WWW 2011, pp. 15–16 (2011)
12. Xue, X., Jeon, J., Croft, W.B.: Retrieval models for question and answer archives. In: Proceedings of the 31st Annual International ACM SIGIR Conference on Research and Development in Information Retrieval, SIGIR 2008, pp. 475–482 (2008)
13. Carpineto, C., Romano, G.: A survey of automatic query expansion in information retrieval. ACM Comput. Surv. **44**(1), 1:1–1:50 (2012). https://dl.acm.org/citation.cfm?id=2071390
14. Bhogal, J., Macfarlane, A., Smith, P.: A review of ontology based query expansion. Inf. Process. Manage. **43**, 866–886 (2007)
15. Buckley, C.: Automatic query expansion using smart: TREC 3. In: Proceedings of The third Text Retrieval Conference, pp. 69–80 (1994)
16. Salton, G., Buckley, C.: Readings in Information Retrieval, pp. 355–364. Morgan Kaufmann Publishers Inc., San Francisco (1997)
17. Rocchio, J.J.: Relevance feedback in information retrieval. In: The Smart Retrieval System - Experiments in Automatic Document Processing, pp. 313–323 (1971)
18. Chen, C., Chunyan, H., Xiaojie, Y.: Relevance feedback fusion via query expansion. In: 2012 IEEE/WIC/ACM International Conferences on Web Intelligence and Intelligent Agent Technology, pp. 122–126 (2012)
19. Wang, Y., Fang, H.: Exploring the query expansion methods for concept based representation. In: The Twenty-Third Text Retrieval Conference Proceedings, TREC 2014 (2014)
20. Xu, J., Croft, W.B.: Query expansion using local and global document analysis. In: Proceedings of the 19th Annual International ACM SIGIR Conference on Research and Development in Information Retrieval, SIGIR 1996, pp. 4–11 (1996)
21. Lv, Y., Zhai, C.: Positional relevance model for pseudo-relevance feedback. In: Proceedings of the 33rd International ACM SIGIR Conference on Research and Development in Information Retrieval. SIGIR 2010, pp. 579–586 (2010)

22. Ermakova, L., Mothe, J., Nikitina, E.: Proximity relevance model for query expansion. In: Proceedings of the 31st Annual ACM Symposium on Applied Computing, SAC 2016, pp. 1054–1059 (2016)
23. Zamani, H., Croft, W.B.: Relevance-based word embedding. In: Proceedings of the 40th International ACM SIGIR Conference on Research and Development in Information Retrieval, SIGIR 2017, pp. 505–514 (2017)
24. Martinez, D., Otegi, A., Soroa, A., Agirre, E.: Improving search over electronic health records using UMLS-based query expansion through random walks. J. Biomed. Inform. **51**, 100–106 (2014)
25. Zhu, D., Carterette, B.: Combining multi-level evidence for medical record retrieval. In: Proceedings of the 2012 International Workshop on Smart Health and Wellbeing (SHB 2012), pp. 49–56 (2012)
26. Wang, Y., Fang, H.: Exploring the query expansion methods for concept based representation. In: TREC 2014 (2014)
27. Limsopatham, N., Macdonald, C., Ounis, I.: Learning to combine representations for medical records search. In: Proceedings of SIGIR 2013 (2013)
28. Wang, Y., Liu, X., Fang, H.: A study of concept-based weighting regularization for medical records search. In: ACL 2014(2014)
29. Zhai, C., Lafferty, J.: A study of smoothing methods for language models applied to information retrieval. ACM Transactions on Information Systems, pp. 179–214 (2004)
30. Lavrenko, V., Croft, W.B.: Relevance based language models. In: Proceedings of the 24th Annual International ACM SIGIR Conference on Research and Development in Information Retrieval. SIGIR 2001, pp. 120–127 (2001)
31. Mikolov, T., Chen, K., Corrado, G., Dean, J.: Efficient estimation of word representations in vector space. Computing Research Repository (2013)

Ad-hoc Video Search Improved by the Word Sense Filtering of Query Terms

Koji Hirakawa[1], Kotaro Kikuchi[1], Kazuya Ueki[1,2], Tetsunori Kobayashi[1], and Yoshihiko Hayashi[1(✉)]

[1] Waseda University, Tokyo, Japan
hirakawa@pcl.cs.waseda.ac.jp, yshk.hayashi@aoni.waseda.jp
[2] Meisei University, Hino, Japan

Abstract. The performances of an ad-hoc video search (AVS) task can only be improved when the video processing for analyzing video contents and the linguistic processing for interpreting natural language queries are nicely combined. Among the several issues associated with this challenging task, this paper particularly focuses on the sense disambiguation/filtering (WSD/WSF) of the terms contained in a search query. We propose WSD/WSF methods which employ distributed sense representations, and discuss their efficacy in improving the performance of an AVS system which makes full use of a large bank of visual concept classifiers. The application of a WSD/WSF method is crucial, as each visual concept classifier is linked with the lexical concept denoted by a word sense. The results are generally promising, outperforming not only a baseline query processing method that only considers the polysemy of a query term but also a strong WSD baseline method.

Keywords: Ad-hoc video search · TRECVID AVS · Visual concepts
Query processing · Word sense disambiguation

1 Introduction

Given a huge amount of video contents available on the Web, an intelligent yet handy method for accessing these contents is highly demanded [3]. With this requisition in the scope, the TRECVID Ad-hoc Video Search (AVS) workshop[1] maintains a technical forum that promotes progress in content-based exploitation of digital videos.

Given an English ad-hoc query like "Find shots of a person playing drum indoors", an AVS system is required to return a ranked list of possibly-relevant video shots from the test collection that collects more than 330k video shots. As suggested by this example, an AVS system has to match the search condition

---

[1] https://trecvid.nist.gov/.

Y.-H. Tseng et al. (Eds.): AIRS 2018, LNCS 11292, pp. 157–163, 2018.
https://doi.org/10.1007/978-3-030-03520-4_15

latent in the query against the visual features extracted from the set of video shots, requiring both the video and query processing work well in tandem.

Among the variety of approaches proposed in the NIST shared tasks [1], we decided to focus on a type of search system which employs a large array of visual concept classifiers. As our primarily goal in the present research is to explore the efficacy of word sense disambiguation/filtering (WSD/WSF) of query terms, this choice of target system type is reasonable: each of the majority of visual concept classifiers are associated with an ImageNet [2] concept, which is, in turn, linked to a synset in the lexical database of English, WordNet [5].

In the reminder of this paper, our base system is first introduced, followed by the proposal of two WSD/WSF methods, each of which is evaluated by implemented in the base system. These methods rely on distributed sense representations derived by a system called AutoExtend [7]. The experimental results using the TRECVID AVS 2016 test collection is finally presented to empirically show the efficacy of the WSD/WSF methods in ad-hoc video search.

2   Overview of the Base System

Figure 1 illustrates the processing pipeline of our base system, which basically is a system participated in the TRECVID AVS shared task [9]. Each term extracted from the given query is matched against WordNet (WN), creating a set of WN synsets[2]. As each WN synset in this set is associated with an ImageNet synset if it is a visual concept, the associated visual concept classifier, as well as other types of non-ImageNet-based classifiers[3], are invoked, and they provide the score s_i for a video shot v_k in the test collection V. Just in case, the associated visual concepts for a query term has not been be found in ImageNet, only the non-ImageNet-originated classifiers associated with the term are invoked.

The total score $S(q, v_k)$ of the video shot v_k given a query q is given by the following formula. Here s_i is the averaged score of the visual concept classifiers invoked for the term i, normalized by min-max normalization that uses the whole test collection. This normalization process is required, as the range of classifier scores varies. We experimentally decided to use averaged scores instead of other types of scores, such maximum scores. Note further that c_i is the IDF weight of the term i, which was computed by using the collection of captions given in an image dataset[4]. These weights are effective in reducing the impact of low-scored classifiers.

$$S(q, v_k) = \prod_{i=1}^{N} s_i^{c_i} \tag{1}$$

---

[2] A WordNet synset denotes a lexical concept. It is defined by a set of synonymous word senses. A word, more precisely a word form, generally has multiple senses and each sense denotes a unique synset.

[3] Refer to [9] for the list of employed classifiers.

[4] We employed the MS COCO dataset available at http://cocodataset.org.

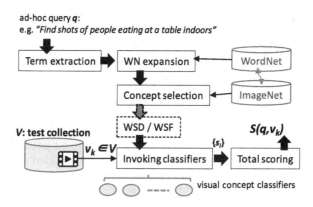

Fig. 1. The organization of the base AVS system.

Selection of Visual Concepts: To consider the polysemy of a word, each term (content word) in a query is matched against WN (henceforth, *WN expansion*), yielding a set of synsets. Of which, some possibly non-visual synsets that are not associated with ImageNet synsets are discarded. Still the synset set may contain synsets that are irrelevant to the sense of a term in the query. These irrelevant synsets could degrade the AVS performances as dictated in Eq. 1. We thus expect that an effective WSD/WSF method could exclude semantically irrelevant concepts, contributing to the improvement in the AVS performances.

3   Word Sense Disambiguation/Filtering

The task of WSD is defined as to select the appropriate sense of a word in a given linguistic context. It has long been a central issue in natural language processing [6], and the research field is still struggling to develop a method that could outperform the most frequent sense (MFS) heuristic, which is highly robust and strong. In general, two fundamental ingredients are necessary in WSD, which are a sense inventory and a computational method to deal with the linguistic concept of a target word.

3.1   Sense Inventory

We utilize WordNet as the sense inventory: it is a de fact standard lexical resource of English, and more importantly, many of the concrete concepts are associated with corresponding ImageNet synsets, meaning that we can obtain nice sets of images for these concepts. This allows us to adopt an AVS approach which employs a *concept base*: a large set of image classifiers trained by the images indexed by visual concepts [9]. Note that the base system accommodates more than 50k classifiers, including some that are not indexed by visual concepts.

3.2 Proposed WSD Methods

We propose two WSD methods, *SimSum* and *DistSim*, both employ sense representations derived by the AutoExtend system [7]. This system generates a set of distributed representations for the word senses (lexemes) and the synsets defined in WordNet by being fed a set of word embedding vectors. One of the nice properties with AutoExtend is that the sense embeddings are created in the same space of the word embeddings, allowing us to compare embeddings of different types. For example, we can easily calculate the relatedness between a word and a WN synset. In the present work, we employed the pre-trained set of Word2Vec [4] word embeddings[5] generated from GoogleNews corpus.

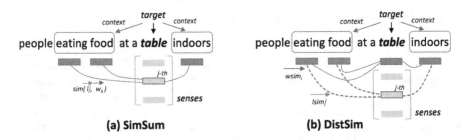

(a) SimSum **(b) DistSim**

Fig. 2. The conceptual illustration of the WSD methods.

Similarity Summation (SimSum) Method): As illustrated in Fig. 2.-(a) and defined by Eq. 2, the sense $\hat{l}_i$ of term w_i in a query is determined by comparing the summed the cosine similarities between a sense w_i^k of w_i and each of the word w_k in $C(w_i)$, which represents the set of words in $2M$-word context of w_i. Note that this method expects that semantically related words may occur within the context of a target word, and hence the correct sense can be selected by looking at the similarities of nearby context words.

$$\hat{l}_i = \arg \max_{j} \sum_{w_k \in C(w_i)} sim(\overrightarrow{l_i^j}, \overrightarrow{w_k}) \tag{2}$$

Distributional Similarity (DistSim) Method: This method, defined by Eq. 3 and illustrated in Fig. 2.-(b), tries to identify the sense $\hat{l}_i$ of term w_i by considering the contextual similarity between two vectors: $\overrightarrow{wsim_i}$ and $\overrightarrow{lsim_i^j}$. The former vector represents the context of w_i by arraying the word-word similarities of $2M$-word window context, whereas the latter vector dictates the same context by assuming that the word token w_i carries its j-th sense in the query context. Notice that

---

[5] https://code.google.com/archive/p/word2vec/.

this method implements the notion of *distributional hypothesis* whose claim is stated as "a word is characterized by the company it keeps".

$$\hat{l}_i = \arg\max_j sim(\overrightarrow{wsim_i}, \overrightarrow{lsim_i^j}) \tag{3}$$

Here, the vectors fed into the similarity function are constructed as follows.

$$\overrightarrow{wsim_i} = (sim(\overrightarrow{w_{i-M}}, \overrightarrow{w_i}), \cdots, sim(\overrightarrow{w_{i-1}}, \overrightarrow{w_i}), sim(\overrightarrow{w_{i+1}}, \overrightarrow{w_i}), \cdots, sim(\overrightarrow{w_{i+M}}, \overrightarrow{w_i}))$$

$$\overrightarrow{lsim_i^j} = (sim(\overrightarrow{w_{i-M}}, \overrightarrow{l_i^j}), \cdots, sim(\overrightarrow{w_{i-1}}, \overrightarrow{l_i^j}), sim(\overrightarrow{w_{i+1}}, \overrightarrow{l_i^j}), \cdots, sim(\overrightarrow{w_{i+M}}, \overrightarrow{l_i^j}))$$

3.3 Word Sense Filtering (WSF)

The division of word senses are somewhat arbitrary, and their correspondences between visual concepts are often not clear-cut. Thus we assess WSD as well as WSF in the experiments. A WSF method allows multiple senses to be remained rather than selects only the best one.

4 Experimental Results and Discussion

4.1 Experimental Settings

We conducted a series of video search experiments by using the TRECVID AVS 2016 test collection, which collects more than 330k video shots. The test collection maintains 30 ad-hoc queries. For each query, the system composes a ranked list of possibly-relevant video shots. We employed a concept bank that accommodates more than 50k visual concept classifiers, each trained by using a relevant image dataset and a visual recognition model. For the context window size used in the similarity computation, we experimentally set $M = 3$.

As the system is required to rank the retrieved video shots for each query, we can compute the average precision (AP), provided that a set of gold video shots is given. The total evaluation metrics employed in the experiments is thus the mean average precision (mAP)[6], which is the mean of APs over all queries.

4.2 Results

Table 1 summarizes the experimental results, where the `Baseline` column displays the mAP result with WordNet expansion, but without any WSD/WSF. The mAP result 0.177 was far better than that of the most primitive setting without WordNet expansion, which was only 0.152. In this primitive setting, the system directly matched each query term with the names of visual concept classifiers, showing the benefit of consulting WordNet.

The `WSD` columns display the mAP results with each of the WSD methods, showing that the MFS heuristic was as good as the DistSim method, but worse

---

[6] The official evaluation metrics adopted by TRECVID AVS is a variant of usual mAP.

Table 1. mAP results by WSD/WSF methods.

Baseline (without WSD/WSF)	WSD			WSF	
	MFS	SimSum	DistSim	SimSum	DistSim
0.177	0.174	0.178	0.174	0.181	0.176

than the `Baseline` method that does not conduct WSD. In contrast, the Sim-
Sum method slightly outperformed these methods. These results indicate that
WSD, in general, is a difficult task, also in the present situation of ad-hoc query
processing[7].

The `WSF` columns compare the two proposed methods in the *filtering* setting
where multiple senses are allowed to be remained: We kept the senses whose
similarity were greater than 70% (SimSum)/50% (DistSim) of the maximum
similarity. The displayed mAP results are better than that of the WSD setting,
insisting that the senses of query terms should be properly narrowed down, but
not necessarily be completely resolved. Also notice that the SimSum method was
again better than the DistSim method, demonstrating that a simple method is
often better than more principled methods.

4.3   Discussion

The differences in the performances of WSD and WSF settings may be worth
explored. They could partly be attributed to the nature of the ImageNet dataset,
in which the linguistic sense divisions may not be well aligned with the visual
concept distinctions. This issue may also be linked to the often discussed problem
of the too-fine-grained WordNet sense divisions [8].

5   Concluding Remarks

This paper argued that the performances of AVS could be improved by properly
considering the senses of a term appeared in a phrasal query. More specifically,
we proposed two WSD/WSF methods relying on distributed sense representations. Although the overall experimental results were promising, several issues
are remained to be addressed. Most importantly, we need to examine the interactions between the accuracy of WSD/WSF and the performances of AVS. This
would also be prerequisite in addressing the issue of synonymy of a concept. To
do this, the senses of words not presented in the query but similar in meaning
have to be incorporated.

---

[7] As the number of queries in the TRECVID AVS task is as small as 30, the WSD
accuracies by the presented methods are quite unstable. We could not observe any
statistical significance. However the DistSim method slightly outperformed two other
methods in precision: 0.892 (DistSim) to 0.890 (MFS) and 0.888 (SimSum).

Acknowledgment. The present work was partially supported by JSPS KAKENHI Grants numbers 15K00249, 17H01831, and 18K11362, and the Kayamori Foundation of Informational Science Advancement.

References

1. Awad, G., et al.: Trecvid 2017: evaluating ad-hoc and instance video search, events detection, video captioning and hyperlinking. In: Proceedings of TRECVID 2017. NIST, USA (2017)
2. Deng, J., Dong, W., Socher, R., jia Li, L., Li, K., Fei-fei, L.: Imagenet: a large-scale hierarchical image database. In: CVPR (2009)
3. Inoue, N., Shinoda, K.: Semantic indexing for large-scale video retrieval. ITE Trans. Media Technol. Appl. **4**(3), 209–217 (2016). https://doi.org/10.3169/mta.4.209
4. Mikolov, T., Sutskever, I., Chen, K., Corrado, G., Dean, J.: Distributed representations of words and phrases and their compositionality. In: Proceedings of the 26th International Conference on Neural Information Processing Systems - Volume 2, NIPS 2013, pp. 3111–3119. Curran Associates Inc., USA (2013)
5. Miller, G.A., Fellbaum, C.: Wordnet then and now. Lang. Resour. Eval. **41**(2), 209–214 (2007). https://doi.org/10.1007/s10579-007-9044-6
6. Navigli, R.: Word sense disambiguation: a survey. ACM Comput. Surv. **41**(2), 1–69 (2009)
7. Rothe, S., Schütze, H.: Autoextend: combining word embeddings with semantic resources. Comput. Linguist. **43**(3), 593–617 (2017)
8. Snow, R., Prakash, S., Jurafsky, D., Ng, A.Y.: Learning to merge word senses. In: Proceedings of the 2007 Joint Conference on Empirical Methods in Natural Language Processing and Computational Natural Language Learning (EMNLP-CoNLL) (2007)
9. Ueki, K., Hirakawa, K., Kikuchi, K., Ogawa, T., Kobayashi, T.: Waseda_meisei at trecvid 2017: Ad-hoc video search. In: 2017 TREC Video Retrieval Evaluation Notebook Papers (2017)

Considering Conversation Scenes in Movie Summarization

Masashi Inoue[1](✉) ⓘ and Ryu Yasuhara[2]

[1] Tohoku Institute of Technology, Yagiyama Kasumicho 35-1, Sendai, Japan
m.inoue@acm.org
[2] Yamagata University, Jyonan 4-3-16, Yonezawa, Japan
http://www.ice.tohtech.ac.jp/~inoue/index.html

Abstract. Given that manual video summarization is time consuming and calls for a high level of expertise, an effective automatic video summarization method is required. Although existing video summarization methods are usable for some videos, when they are applied to story-oriented videos such as movies, it sometimes becomes difficult to understand the stories from the generated summaries because they often lack continuity. In this paper, we propose a method for summarizing videos that can convey the story beyond the sequence of extracted shots so that they can fit user perception patterns. In particular, we examine the impact of conversation scenes in movie storytelling. The evaluation of summarized videos is another challenge because existing evaluation methods for text summarization cannot be directly applied to video summarization. Therefore, we propose a method for comparing summarized movies that maintains the integrity of conversation scenes with those that do not. We demonstrate how preserving conversational aspects influences the quality of summarized videos.

Keywords: Video summarization · Movie summarization
Evaluation · Storytelling · Conversation

1 Introduction

A summary video presents the important parts of a video usually by combining short video segments extracted from the original video. However, it is difficult and time consuming to prepare a summary video manually. To address this problem, various automatic summarization methods have been studied [6]. Among the various types of video that exist, it is relatively easy to generate a summary video where the contents are stylized and have a less story-oriented nature, such as sports videos in which redundant and highlighted sections are identifiable through machine-processable, low-level features. In the case of story-oriented videos such as movies and dramas, it is difficult to determine the important sections for generating a relevant summary using a computer.

© Springer Nature Switzerland AG 2018
Y.-H. Tseng et al. (Eds.): AIRS 2018, LNCS 11292, pp. 164–170, 2018.
https://doi.org/10.1007/978-3-030-03520-4_16

In this research, we aim to provide a method for automatically generating a summary video for story-oriented videos for the purpose of increasing understanding and enjoying. Movies and dramas tell a story, but it is unclear which sections in the video are involved in the progress of the story. In fact, important segments are difficult to determine based on low-level features alone such as audio and visual information. Therefore, there are semantic video summarization methods proposed. The difficulty in semantic summarization is that the model used for representing deep semantics are often complex and obtaining high-level feature is costly when they are created manually. Therefore, it is desirable to estimate some shallow semantic features from from low-level features. As a computationally derivable shallow semantic feature, we focus on conversation scenes. Although conversation scenes were used for video abstraction [3], the evaluation of the generated summary videos remains as a major problem. Therefore, we proposed a text description-based and crowdsourcing method for quantitative evaluation. The role of conversation in movie for effective summarization is clarified through the experiment.

2 Materials

There are few story-oriented video data sets that can be used for the evaluation of video summarization owing to copyright restrictions. In terms of data for the experiment, accessible data with Creative Commons (CC) or Public Domain (PD) licenses are desirable for their reproducibility and usability.

Therefore, in this research, we use a publicly available data set that is a collection of public domain movies. In addition to the video files, title text, and video description text, we utilized automatically assigned conversation section information and genre information provided in the original data set [7][1]. The videos are movies with a CC license that are hosted on Internet Archive[2]. The genre information is given as 22 genre tags defined in the Internet Movie Database (IMDb)[3].

All $1,722$ movies in the dataset were plotted based on both the frequencies of utterances in the movie and on the average duration of utterances (Fig. 1). We defined three groups as the three clusters found after plotting. The groups were (G1) with many conversations, an intermediate group (G2), and a group with few conversations (G3). We selected five works for evaluation from each of the three groups of (Table 1).

3 Method

3.1 Base Process

Video Segmentation. The summarization is carried out in three steps: video segmentation, feature extraction and importance assignment, and summary

---

[1] http://www.ice.tohtech.ac.jp/~inoue/moviedialcorpus/index.html.
[2] https://archive.org/index.php.
[3] http://www.imdb.com/.

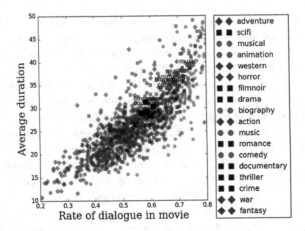

Fig. 1. Distribution of utterance frequencies and utterance duration.

Table 1. List of target movies.

Group	Title	Genre	Duration (min:sec)	Task ID
G1	Dreaming Out Loud	Comedy	65:10	1
	Hey! Hey! USA	Comedy	87:56	2
	The Ghost Walks	Comedy	63:26	3
	Windbag the Sailor	Comedy	81:30	4
	Texas, Brooklyn and Heaven	Comedy	76:18	5
G2	Cosmos: War of the Planets	Sci-Fi	89:03	1
	The Great Commandment	Drama	80:12	2
	First Spaceship on Venus	Sci-Fi	78:31	3
	A Star Is Born	Drama	110:53	4
	Night Alarm	Drama	61:08	5
G2	Svengali	Horror	81:08	1
	Under California Stars	Western	72:20	2
	Hollywood Man	Action	84:34	3
	Pecos Kid	Western	54:09	4
	Night of the Living Dead	Horror	95.52	5

video generation. In the first step, there are differences in the granularity of divisions such as frame, shot, scene, and sequence. Among them, we used the shot unit as a video section in which a single camera shoots consecutively. For segmentation, we used PySceneDetect[4] as a tool for detecting a shot change by observing the change in the amount of the difference of HSV histograms between frames. The precision and recall values were 0.885 and 0.830 respectively.

---

[4] https://github.com/Breakthrough/PySceneDetect.

Importance Scoring. In order to select the shots included in the summary from the segmented video, the importance is calculated for each shot. Ma et al., combined low-level features to determine importance scores to estimate people perception at a higher cognitive level [4]. We adopt their method and leverage a model that estimates the part that attracts people's interest. The method of Ma et al. does not target videos with a narrative nature, but their model is considered useful for wide variety of videos.

Video Generation. The selection of the video segments to be included in the summary can be considered as a 0–1 knapsack problem selecting shots that maximize the obtained importance score so that it fits within the limited duration of the post-summarized video. In this research, a summarized video is generated with a dynamic programming algorithm [5].

3.2   Conversation Integration

When a conversation scene is divided into shots as the basis for the summary video, and if the shot in the middle of the scene is not rated as important, there is the possibility that the information during the conversation drops out, making it difficult to understand the contents. Therefore, the proposed method explicitly uses the conversation section information and the divided shots are grouped into conversation units. For example, for a scene S consisting of shots $S = \{s_1, s_2, ..., s_5, s_6, ..., s_n\}$, if it is assumed that three cut points in the second to fifth shots are within the conversation section, $s_2, ..., s_5$ are merged and the set of shots after the integration are $S' = \{s_1, \hat{s_1}, s_6, ..., s_n\}$ where $\hat{s_1} = \{s_2, s_3, s_4, s_5\}$.

4   Evaluation

4.1   Evaluation Procedure

An evaluation of the video summary is performed via a subjective evaluation method that shows the automatically generated videos to the human assessors and collects their evaluations. The evaluation indices consist of: informativeness (information quality) and enjoyability (entertainment quality). Informativeness is an index which shows how much the summary video preserved the information necessary for understanding the content compared to the original video, and enjoyability is an index which indicates the degree of satisfaction with the summary video [4].

In order to measure the informativeness, it is necessary for the subject to know the contents of the original video. If the assessors have already seen the movies, they can use their knowledge of the respective films. However, the movies collected in this research consist of many old or less-known releases, and it is unlikely that the participants have watched them before the task. In addition, considering the movie running time, the burden of viewing the summary video after viewing the original video becomes too large. Therefore, in this research, we provided text information explaining the outline of the movie to enable users to grasp the content of the original movie after watching summaries. By comparing the information given by watching summarized video and the text explanation that is considered as the ground truth, it becomes possible to estimate the degree of informativeness of the summaries. For the text describing the outline of the movie to be used, the story-line (outline) information of the movie has been taken from the IMDb movie database where the information is created by user postings. Conversely, enjoyability can be judged without any additional information. Assessors were asked if they enjoyed the summary video they just watched as a video, and they provided a score in 10 steps. Because the target movies are in English, we collected assessors who understand English using a crowdsourcing service (CrowdFlower[5]). There are five video set (tasks) and 10 assessors were assigned for each task. That is, 50 assessors participated in total. In addition to the scores of informativeness and enjoyability, the subjects were asked to answer questions on the summary videos in complete sentences so that they could be used for analysis. An answer was collected by setting free description columns asking what type of movie it was, what type of information was lacking to understand the content, and why they could or could not enjoy the video.

Each participant watched only one of the videos generated by the proposed method and that by the conventional method. In other words, each participant responded to three works in total, one for each crowdsourcing task ID from each group (G1, G2, or G3). Ten crowdworkers participated in each task (evaluation of three summaries) and each group assigned 50 scores for both evaluation measures.

4.2   Results

Table 2 shows the average scores of informativeness and enjoyability for each summary. When considering conversation, the average score of informativeness improved by 0.56 points, 0.30 points, and 0.48 points, respectively, for each group compared with the case that did not consider conversation. Likewise, the average enjoyability score improved by 0.74 points, 0.58 points, and 0.68 points, respectively, for each group. When individual films were considered, the results were mixed. From these results, it can be predicted that it became easier to understand the contents of the movie by generating the summary video grouping the conversational shots, which also confirmed that the naturalness of the video can also be preserved.

Additionally, by analyzing the textual responses provided by the assessors, we were able to determine the qualitative differences among the two methods. One of them is the usage of proper nouns. A proper noun appearing in the video such as the name of a character or a place name is considered to be an important element for understanding and explaining the contents. By considering the method that did not leverage the conversations, the explanation using a proper noun in the video increased 6.0%. As an example, if a participant watched a summary video that did not consider conversations, the explanation

---

[5] https://www.crowdflower.com.

used nouns like "reporter goes to NY," whereas in the text explaining the summary video that considered conversation descriptions could be found similar to: "Eddie Taylor leaves Dallas, Texas and his newspaper job with an inheritance for New York." Therefore, explanations using proper nouns seem to be more concrete. Another example shows that without considering conversation consistency in the summary video, the description is plain: "Zombies start attacking a girl and kill the guy she's with." Conversely, if conversation units are taken into account, participants are able to understand the relationships between characters: "Barbara and her brother Johnny decided to visit their parents' grave."

Table 2. Result of human assessments on a ten-point scale.

		Group 1		Group 2		Group 3	
		Info	Enjoy	Info	Enjoy	Info	Enjoy
1	Considering conversation	5.0	4.8	4.1	4.2	4.4	4.3
	Not considering conversation	5.2	4.1	4.8	2.9	5.3	4.0
2	Considering conversation	6.2	5.9	6.2	4.4	5.8	5.8
	Not considering conversation	4.9	3.5	6.4	5.5	6.0	5.0
3	Considering conversation	7.6	6.7	7.4	6.4	7.0	5.1
	Not considering conversation	5.6	5.5	5.4	4.2	4.3	4.1
4	Considering conversation	7.4	6.3	7.4	7.0	6.8	6.1
	Not considering conversation	7.7	6.4	6.8	5.7	6.2	5.0
5	Considering conversation	6.7	5.4	6.7	5.7	8.1	7.9
	Not considering conversation	6.7	5.9	6.9	6.5	7.9	6.7
Average	Considering Conversation	6.58	5.82	6.36	5.54	6.42	5.84
	Not considering conversation	6.02	5.08	6.06	4.96	5.94	4.96

5 Conclusion

In this research, we proposed a method for the automatic generation of a summarized video based on a story-oriented video such as a movie. In the proposed method, the continuity of the information is maintained by preserving the conversation segments in a summary video to achieve semantic cohesion. Owing to the fact that the automatic evaluation of summarized videos is difficult and human evaluation by comparing original and summarized videos is time consuming, we proposed a text description-based and crowdsourcing methods for summary video evaluation. As a result, a subjective comparison by crowd assessors of summarized videos showed that the proposed method was rated as superior in terms of informativeness and enjoyability. The compression rate of the summarized video was about 30% in our setting. The influence of compression rate should be investigated.

Future topics include the expansion of evaluation measures and improvement of the conversation scene extraction. The evaluation measures used in this research are informativeness and enjoyability. In the task of video search, based on a user study, there are 28 evaluation criteria suggested [1]. Their applicability to the summarization task can be considered. Regarding the conversation scene detection, improvements of VAD algorithm that was applied for creating the dataset used in this study by incorporating additional noise classes [2]. Investigation of the improved conversation scene information in summarization quality is an interesting future work.

References

1. Albassam, S.A.A., Ruthven, I.: Users' relevance criteria for video in leisure contexts. J. Doc. **74**(1), 62–79 (2018)
2. Kosaka, T., Suga, I., Inoue, M.: Improving voice activity detection for multimodal movie dialogue corpus. In: IEEE 7th Global Conference on Consumer Electronics (GCCE 2018), Nara, Japan, pp. 453–456 (2018)
3. Lienhart, R., Pfeiffer, S., Effelsberg, W.: Video abstracting. Commun. ACM **40**(12), 54–62 (1997)
4. Ma, Y.F., Lu, L., Zhang, H.J., Li, M.: A user attention model for video summarization. In: Proceedings of the Tenth ACM International Conference on Multimedia, MULTIMEDIA 2002, Juan-les-Pins, France, pp. 533–542 (2002)
5. McDonald, R.: A study of global inference algorithms in multi-document summarization. In: Amati, G., Carpineto, C., Romano, G. (eds.) ECIR 2007. LNCS, vol. 4425, pp. 557–564. Springer, Heidelberg (2007). https://doi.org/10.1007/978-3-540-71496-5_51
6. Truong, B.T., Venkatesh, S.: Video abstraction: a systematic review and classification. ACM Trans. Multimed. Comput. Commun. Appl. **3**(1), 1–37 (2007)
7. Yasuhara, R., Inoue, M., Suga, I., Kosaka, T.: Large-scale multimodal movie dialogue corpus. In: Proceedings of the 18th ACM International Conference on Multimodal Interaction, ICMI 2016, Tokyo, Japan, pp. 414–415 (2016)

Best Paper Session

Hierarchical Attention Network for Context-Aware Query Suggestion

Xiangsheng Li[1], Yiqun Liu[1(✉)], Xin Li[1], Cheng Luo[1], Jian-Yun Nie[2], Min Zhang[1], and Shaoping Ma[1]

[1] Department of Computer Science and Technology,
Institute for Artificial Intelligence,
Beijing National Research Center for Information Science and Technology,
Tsinghua University, Beijing 100084, China
yiqunliu@tsinghua.edu.cn
[2] Université de Montréal, Montreal, Canada

Abstract. Query suggestion helps search users to efficiently express their information needs and has attracted many studies. Among the different kinds of factors that help improve query suggestion performance, user behavior information is commonly used because user's information needs are implicitly expressed in their behavior log. However, most existing approaches focus on the exploration of previously issued queries without taking the content of clicked documents into consideration. Since many search queries are short, vague and sometimes ambiguous, these existing solutions suffer from user intent mismatch. To articulate user's complex information needs behind the queries, we propose a hierarchical attention network which models users' entire search interaction process for query suggestion. It is found that by incorporating the content of clicked documents, our model can suggest better queries which satisfy users' information needs. Moreover, two levels of attention mechanisms are adopted at both word-level and session-level, which enable it to attend to important content when inferring user information needs. Experimental results based on a large-scale query log from a commercial search engine demonstrate the effectiveness of the proposed framework. In addition, the visualization of the attention layers also illustrates that informative words and important queries can be captured.

Keywords: Query suggestion · Recurrent neural networks
Click-through behavior

1 Introduction

Web search queries are usually short and ambiguous [1]. According to the investigations conducted on large-scale commercial search engines, a query often contains less than 3 terms [2] and over 16% of queries are ambiguous [3]. It is therefore challenging for search engines to understand user's search intents. Query suggestion is widely applied by commercial search engines to help users organize

© Springer Nature Switzerland AG 2018
Y.-H. Tseng et al. (Eds.): AIRS 2018, LNCS 11292, pp. 173–186, 2018.
https://doi.org/10.1007/978-3-030-03520-4_17

their queries and express their information needs more efficiently. It is shown that query suggestion can significantly improve user satisfaction, especially for informational queries [4].

Existing approaches for query suggestion mainly focus on mining the co-occurred queries from query log [5,6]. The assumption is that a frequently co-occurred query is likely to be the next query issued by users. These methods suffer from data sparsity and can hardly provide satisfactory results for long-tail queries [7]. To handle this problem, more features based on click-through (e.g., click position, click frequency and dwell time) are exploited [8,9]. They assumed that different users share common interests with each other when their click behaviors are similar. However, this assumption is not reasonable in many cases since the issued queries are ambiguous. The key problem is to improve the query representation with more clear search intent.

Liu et al. [10] analyzed user's search processes and concluded that in addition to the previous queries, user's search intent can also be reflected by the clicked results. We follow this observation and attempt to incorporate the entire search interactions into query suggestion. Due to the brevity of query, precisely expressing user information needs in queries is intractable. Clicked documents can be regarded as an implicit description of the issued query and enable us to better infer user's information needs. For example, a user submits a query "Tourism" to the search engine, and clicks a document about "Shopping centers". It is intuitive to suggest queries about tourism if we only consider the information of query. But the clicked document reflects that this user is more interested in shopping during the traveling. Existing methods do not consider this information and thus ignore user's diverse and detailed search preference. If we can take advantage of this observation, the suggested query will be closer to the real information needs. Using clicked documents can be helpful to infer the precise search intent behind the short queries.

In this paper, we propose a hierarchical attention network (HAN) which models not only the issued queries but also the clicked documents in a whole session. The network contains three layers of encoder-decoder based on recurrent neural networks (RNN). The first layer concatenates two encoders which model the clicked documents and queries, respectively. Then a session-level encoder summarizes the information of previous search process. The final decoder is utilized to predict the next query according to the session information encoded by the session-level encoder. On the other hand, context information in a session generally has different importance because only a few words contribute to the inference of information needs. To capture the pivot information automatically, we construct two levels of attention mechanisms at word-level and session-level. The word-level attention enables us to focus on informative and important words in clicked documents and queries while the session-level attention aims to recognize the useful queries (e.g., queries with informative words or click feedbacks).

To validate the effectiveness of our method, we perform our experiments on a query log from a commercial search engine. Comparing to the baselines, our method can significantly improve the performance of query suggestion. Furthermore, we visualize two attention layers and find that the informative words and useful queries in a session can be qualitatively selected. The main contributions of this paper are three-folds:

1. The proposed HAN model encodes both issued queries and the content of clicked documents, which helps us better understanding user information needs.
2. The pivot words and queries in a session can be automatically captured using the attention mechanism without manually selecting pivots.
3. Experiments studies on real-world data show that our model outperforms other baselines on query suggestion.

The reminder of this paper is organized as follows. We review related research studies and compare these work with our approaches in the Sect. 2. In Sect. 3, we detail our proposed hierarchical frameworks. Experiments and technical analysis of our models are reported in Sect. 4. Finally, we conclude this study and highlight the directions of future research work in Sect. 5.

2   Related Work

There have been several studies investigating query suggestion with respect to different search behaviors. Query co-occurrence [11] and term association patterns [12] are common signals used in query suggestion. He et al. [5] proposed a context-aware model called Variable Memory Markov model (QVMM), which builds a suffix tree to model user query sequence. These approaches assume that users share similar search intents with other users who issue similar queries and simply provide query suggestion based on some similarity measures. To provide better query suggestion, click-through behavior [10] is employed along with the issued queries. Liao et al. [13] applied click-through data to build the bipartite graph and clustered queries according to the connections. Jiang et al. [14] exploited query reformulation features to learn users' search behavior and showed the effectiveness for query auto-completion. Different from these methods, we look into users' entire search interaction process by exploiting not only previous queries but also their clicked results, which better satisfy users' search intents.

Other related studies looked into the features resulting from Web search environment. Behaviors such as mouse movements, page scrolls and paginations can also help boost Web search [15,16]. Zhou et al. [17] aggregated user browsing activities for anchor texts, which improved the performance of Web search. Sun et al. [18] focused on right-click query that is submitted to a search engine by selecting a text string in a Web page and extract the contextual information from the source document to improve search results. These studies show the feasibility of search behavior to improve the performance of Web search.

Joachims et al. [19] applied eye-tracking to analyze users' decision processes in Web search and compared implicit feedback against manual relevance judgments. They concluded that users' clicked documents contained valuable implicit feedback information. Sordoni et al. [2] proposed a hierarchical neural networks for query suggestion but only utilizing users' previous queries. Therefore, we look into the content of users' clicked results and incorporate these results into a hierarchical neural model to mine users' information needs.

3   Models

In this section, we introduce the proposed HAN model for context-aware query suggestion. We first give the problem and notations. Then we present our framework of the HAN model, which consists of two attention layers on word-level and session-level, respectively. Finally, we present the details of different components as well as the training process.

3.1   Problem Definition and notations

We regard query suggestion as a sequential encoding and decoding process. A query session S is considered as a sequence of M queries. For each query $Q_m \in S$, it is followed by a sequence of clicked documents $D_m = \{d_1, \ldots, d_n\}$ chronologically. Each of query Q_m and documents d_n consists of a sequence of words w. The task of context-aware query suggestion is to predict the next query Q_m given the context $Q_1, \ldots, Q_{m-1}$ and their clicked documents $D_1, \ldots, D_{m-1}$. Specifically, we predict the query Q_m by reranking a set of candidate queries based on the predicted ranking scores, which is followed by [2]. V is the size of the vocabulary.

3.2   Hierarchical Attention Networks

User preference on search results reflects user's fine-grained information needs. It is proven to be a useful resource in many IR tasks [18,20]. To take advantage of user's search behaviors, we propose a hierarchical attention network (HAN), which models the entire search interactions with search engine as shown in Fig. 1. Specifically, HAN encodes at query-level and session-level hierarchically to model user's information needs. For query-level encoding, since words contribute unevenly to the representation of query embedding, word-level attention mechanism aims to discover the informative words which can best represent the information needs of the current query. For session-level encoding, due to the noisy query in a session [2], the previous query may not be the best query to reflect user information needs in the whole search process. We adopt the session-level attention mechanism to distinguish the difference of the issued queries. In the following, we will detail how we build the session embedding progressively from word embedding by using the hierarchical structure and utilize it to predict the next query.

Query Encoding: For each query $Q_m = \{w_{m,1}, \ldots, w_{m,N_m}\} \in S$ in a session, the content of the responding clicked documents is represented as an aggregation of words $D_m = \{d_1, \ldots, d_n\} = \{w'_{m,1}, \ldots, w'_{m,K_m}\}$, where N_m and K_m are the number of words in the query and the corresponding clicked documents. The

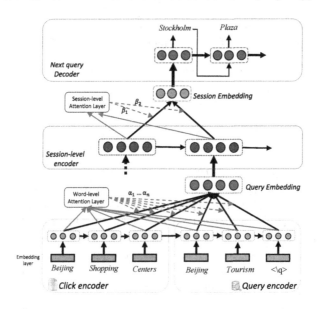

Fig. 1. The architecture of hierarchical attention network for query suggestion. (HAN)

clicked documents under the same query are concatenated chronologically to form a pseudo document of length K_m. According to the statistic of our dataset, a query generates 1.46 clicks on average and 45.45% queries are submitted without any interactions. We then adopt the variant recurrent neural network called gated recurrent unit (GRU) [21] to learn word representation. In particular, query and clicked document are encoded by a query GRU and a click GRU, respectively. If there is no click following the query ($K_m = 0$), click encoder is idle. The representation is obtained by summarizing the contextual information as follows:

$$h_{m,n}^c = GRU_c(h_{m,n-1}^c, w_{m,n}'), n = 1, ..., K_m$$
$$h_{m,n}^q = GRU_q(h_{m,n-1}^q, w_{m,n}), n = 1, ..., N_m$$

$$(1)$$

where $h_{m,n}^c \in \mathbb{R}^{d_h}$ and $h_{m,n}^q \in \mathbb{R}^{d_h}$ are the output recurrent state of click GRU and query GRU, respectively. Click GRU is cascaded to query GRU as in [22], i.e., the initial recurrent state $h_{m,0}^c = 0$ and $h_{m,0}^q = h_{m,K_m}^c$. Each recurrent state stores the order-sensitive information to that position. Then we introduce a word-level attention mechanism to extract the informative words that are important to express the information needs of current query.

Word-Level Attention: A document is much longer than a query, and many words in it are not useful to inferring user information needs. An example is shown in Fig. 1 the word *"Shopping"* is the most important word from the clicked document to understand the information need behind the current query *"Beijing Tourism"*. Thus if we can suggest the next query *"Beijing Plaza"*, which may better express what the user is interested in, thus a possible next query that the user will use. These informative words from the clicked documents are aggregated in the representation of current query embedding as follows:

$$l_{m,t} = sigmoid(W_w h_{m,t} + b_w)$$

$$\alpha_{m,t} = \frac{exp(l_{m,t})}{\sum_j exp(l_{m,j})} \tag{2}$$

$$q_m = \sum_t \alpha_{m,t} h_{m,t}$$

where click recurrent state $h_m^c \in \mathbb{R}^{K_m \times d_h}$ and query recurrent state $h_m^q \in \mathbb{R}^{N_m \times d_h}$ are concatenated into $h_m = [h_m^c, h_m^q] \in \mathbb{R}^{(K_m+N_m) \times d_h}$. The attention weights are estimated by feeding the recurrent state h_m into a one-layer network to get $l_{m,t}$ and further normalized through a softmax function. Finally, we calculate the query embedding as a weighted sum of each click and query recurrent state and use this vector to represent the information needs of current query behavior. The attention layer is learned end-to-end and gradually assign more attention to reliable and informative words.

Session-Level Encoding: Inferring user information needs requires us to consider the whole search behaviors in a session. The previously issued queries and clicked results are both useful to infer user search intents. Therefore, we adopt a session-level GRU, which encodes the previous query embedding to current position:

$$h_m^s = GRU_s(h_m^s, q_m), m = 1, ..., M \tag{3}$$

Each recurrent state $h_m^s \in \mathbb{R}^{d_s}$ incorporates the information of both its surrounding context queries and itself. Then, we apply a session-level attention mechanism to select important queries in the context.

Session-Level Attention: It has been found that a query session may contain queries that are not strongly related to the user's information need, which is called noisy queries (e.g. a user may wonder around in a session) [2]. Therefore, it is necessary to use the previous queries selectively. The importance of queries should be measured correctly in order to better understand user information needs in a session. To do this, we adopt the session-level attention mechanism, which assigns to each query embedding with an attention value.

$$g_{m,t} = sigmoid(W_s h_{m,t}^s + b_s)$$

$$\beta_{m,t} = \frac{exp(g_{m,t})}{\sum_j exp(g_{m,j})} \tag{4}$$

$$s_m = \sum_t \beta_{m,t} h_{m,t}^s$$

Similarly, by combining session recurrent state h_m^s with a one-layer network and normalizing them by a softmax function, we obtain an attention vector β and further an aggregated session embedding $s_m \in \mathbb{R}^{d_s}$, which summarizes the context queries.

Decoding: As shown in Fig. 1, the next query decoder decodes the session embedding s_m to produce the next candidate query. First, the session embedding s_m is transformed to the initial state of the decoder:

$$d_{m,0} = tanh(Ds_m + b) \tag{5}$$

where $|D| = |d_{m,0}| \times |s_m|$ is a projection matrix and b is the bias. The words of next query are decoded by another GRU:

$$d_m = GRU_{dec}(d_{m,n-1}, w_n), n = 1, ..., N_{m+1} \tag{6}$$

and the probability of the next word is:

$$p(w_n|w_{1:n-1}, S) = softmax(w_n f(d_{m,n-1}, w_{n-1})), \\ f(d_{m,n-1}, w_{n-1}) = Hd_{m,n-1} + Ew_{n-1} + b_o \tag{7}$$

where f is a dense layer with parameters H, E and b_o. The softmax layer loops over the vocabulary size V to find next possible word.

In our experiments, we construct a candidate query set for each session and rerank these queries instead of generating the next query directly. The score of a candidate query Q is the probability of being decoded given the session context S:

$$s(Q) = \sum_n \log p(w_n|w_{1:n-1}, S) \tag{8}$$

We utilize this score as an additional feature to combine with a learning-to-rank algorithm to evaluate the reranking performance. It will also be compared with the scores from Sordoni et al. [2], which only models previously issued queries.

This part is similar to the approach of Sordoni et al. [2], except that s_m is enriched with clicked feedback and two levels attention weighting. We will see in our experiments that these additions help to produce better suggestions.

Model Training: Our model is trained end-to-end by using the whole sessions in the query log. Given the issued queries $Q_{1:M}$ and corresponding sets of clicked documents $D_{1:M}$ in a session, each query Q_m is treated as the ground truth based on the context $Q_{1:m-1}$ and $D_{1:m-1}$. The training is conducted by maximizing the log-likelihood of a session S:

$$\mathcal{L}(S) = \sum_m^M \sum_{w_n \in Q_m} \log p(w_n|w_{1:n-1}, S) \tag{9}$$

Table 1. Statistics of the dataset in our experiments.

Dataset	Background	Train	Valid	Test
# Sessions	13,877,582	6,938,508	1,734,596	1,730,773
# Queries	24,572,310	12,286,783	3,070,747	3,062,958

4   Experiments

In this section, we empirically evaluate the performance of the proposed HAN model.

4.1   Dataset

We conduct experiments on the query log from a popular commercial search engine, each entry of which is made up of user ID, issued query, document titles of clicked URLs, and timestamps. Since we are not able to obtain the body text of each document, we only consider the title as the document content. Publicly available query log, e.g., AOL [2], do not contain the detailed content thus are not suitable for our experiment. Queries are split into sessions based on 30 min gap. We shuffle and split them into background, training, validation, and test set with the ratio of 8:4:1:1. The detailed statistics of the dataset is listed in Table 1.

The background set is used to train our model and generate baseline features for a learning-to-rank framework, which follows the prior work [2]. The ranker with only the baseline features is considered as a **Base ranker**. The candidate queries to be reranked are the top 20 most frequent queries based on the co-occurrence with the input query sequence in the background set. A ranking by co-occurrence frequency turned out to be a strong baseline [2]. We call this method the Most Popular Suggestions (**MPS**). Finally, the likelihood score derived from our model is used as an additional feature to produce a new ranker. The likelihood score derived from the baseline **HRED** [2], which only models the previously issued queries, is also used as an additional feature and compared with our model.

We compare the effectiveness of this ranker with other rankers over the training, validation and test set. In testing, we take the last query Q_M and the prior context $Q_{1:M-1}$ and $D_{1:M-1}$ as the ground truth and inputs, respectively. The metric to evaluate the performance of query suggestion is Mean Reciprocal Rank (MRR).

4.2   Experiment Setup

To make a fair comparison of our model with the baseline HRED [2], we use the same parameters as HRED for the common RNN architectures. The dimensionality of query encoder, click encoder, session-level encoder and decoder are set at 300, 300, 600 and 300, respectively. The most frequent $90K$ words in the background set form our vocabulary V. The word embedding, with a dimensionality of 256, is randomly initialized by a standard Gaussian distribution and is trainable during the training. We apply *Adam* to optimize the parameters with mini-batch size 40. The gradients are normalized if their norm exceeds a threshold 1 to stabilize the learning. Early stopping on the validation set is performed during the training process.

LambdaMART is employed as our learning-to-rank algorithm, which is a state-of-the-art supervised ranker that won the Yahoo! Learning to Rank Challenge (2010) [23]. We used the default setting for LambdaMART's prior parameters and the parameters are learned using standard separate training and validation set. The details of baseline features (17 in total) used to train the baseline ranker are listed as follows:

1. Features that describe each suggestion: the suggestion frequency in the background set and the length of the suggestion in terms of number of words and characters.
2. Features that describe the anchor query: frequency of the anchor query in the background set, the times that the suggestion follows the anchor query in the background set and the Levenshtein distance between the anchor query and the suggestion.
3. Features that describe the whole session: character n-gram similarity between the suggestion and 10 most recent queries in the context, the average Levenshtein distance between the suggestion and each query in the context and the estimated scores using the context-aware Query Variable Markov Model (QVMM) [5].

Table 2. The MRR performance of the method. * indicates the statistical significant improvements over each of the baselines.

#	Model	MRR@3	MRR@5	MRR@20
1	MPS	0.5475	0.5678	0.5893
2	Base Ranker	0.5831	0.6077	0.6265
3	+ HRED	0.5913	0.6175	0.6349
4	+ HAN	**0.6042***	**0.6289***	**0.6475***

4.3   Overall Accuracy

Table 2 shows the overall performance of our model and the baselines. In addition to MPS and Base Ranker, we also compare our model with HRED [2], which is similar to ours, except that our session embedding is enriched by user click feedback and two levels of attention weighting. It is observed that the proposed HAN model consistently achieves the best performance on MRR at top 3, 5 and 20.

The improvement due to the addition of a hierarchical encoder-decoder can be seen by comparing HRED and HAN to the base ranker. This result is consistent with [2]. The comparison between HAN and HRED is particularly interesting. The difference between them is due to the utilization of the content of clicked document and to the attention mechanism. We can see that these elements contributed in improving the suggestions. According to our statistics, about 7.4% queries contain words in previous clicked documents but not in previous queries. Globally, query suggestion incorporating clicked documents is more effective and precise.

In addition, our model can automatically capture the pivot information with hierarchical attention mechanism. We observe that most of the words in clicked documents are duplicate or even useless to infer user information needs. Attention mechanism enables us to distinguish these words and assign higher weights to those contributing to the next query. Similarly, the last query is not necessarily the most important nor related query to user information needs. Our model is able to capture the important queries and predict the next query. We will show a detailed case study in Sect. 4.6.

4.4 Effectiveness of Session Length

In order to investigate the effect of session length on our context-aware model, we separate the test set into three categories (The proportion is reported in the brackets):

1. **Short:** Sessions with only 2 queries. (47%)
2. **Medium:** Sessions with 3–4 queries. (34%)
3. **Long:** Sessions with more than 4 queries. (17%)

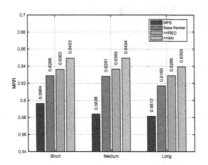

Fig. 2. Performance on sessions with different lengths.

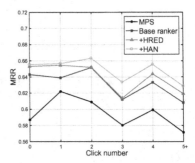

Fig. 3. Performance on sessions with different click frequency.

In Fig. 2, we report the results for different session lengths. We observe that the proposed HAN model outperforms the baselines over different session lengths. As the session becomes longer, MPS performs worse. It is because context information becomes more important, frequency-based method suffers from the noise and sparse signals on longer sessions. Except for MPS, when a model is used for short and medium sessions, its performance is similar. However, we generally observe a lower performance on long sessions. We explain the decrease in performance on long sessions by the fact that the sessions contain more noise. Indeed, in a long session, the searches of the user may be less focused on a specific information need, and queries about unrelated topics could appear. This will create additional difficulties for any methods. The difference in session length does not affect the comparison of our model with the others: In all the three groups of sessions, our model outperforms the others in a similar way.

Table 3. Examples of HAN's query suggestions. The bold words reflect the information needs behind the queries. The purple color represents the session-level attention while the blue color represents the word-level attention. Deeper color implies larger attention weights.

Context 1	Context 2
Q_1 : Food helps to loss weight	Q_1 : How to recover sight?
$C_{1,1}$: Diet to lose weight	$C_{1,1}$: LASIK surgery
Q_2 : Chinese medical cosmetology	Q_2 : Bad Sight
$C_{2,1}$: Chinese medical cosmetology **hospital**	Q_3 : LASIK surgery
	$C_{3,1}$: **HongKong** LASIK hospital
Suggested queries	
1: Plastic surgery **hospital**	1. **HongKong** LASIK surgery
2: Beauty Health	2. **HongKong** LASIK surgery hospital
3: Skin Beauty	3. What if LASIK surgery fails

4.5   Effectiveness of Click Feedback

This experiment further evaluates the effectiveness of click feedback to our model. We split the test set into six categories according to the click frequency in a session, i.e., 0 to 4 and more than 5. Their proportion are 12%, 10%, 16%, 13%, 14% and 35%, respectively. The result is shown in Fig. 3.

In Fig. 3, we observe that a ranker outperforms MPS on all different click frequencies. HRED still outperforms base ranker, but the differences are marginal on frequencies 2 and 3. The proposed HAN model outperforms the base ranker with a quite large margin on all the frequencies. Compared to HRED, HAN produces only marginal improvements when there are limited click information (frequencies 0 and 1). From frequency 2, we can see larger differences between them. This indicates that HAN can benefit more when more click information is available. However, when query sessions become very long, HAN also faces more difficulties to determining what could be the next query due to the problem of noise we discussed in Sect. 4.4. More research is required to infer the topic of the next query from a noisy history.

4.6   Case Study

To better illustrate the effectiveness of our model, we provide two example sessions in Table 3. Based on the context, we predict the next query using standard word-level decoding techniques such as beam-search [24]. This method is able to obtain a predetermined number of best partial suggestions. We list top 3 suggested queries in Table 3.

It is observed that by incorporating the clicked documents, our model can better understand the information needs behind the queries. In *Context* 1, we can only understand that this user is going to learn about beauty by the issued queries. However, the responding clicked document reflects more detailed information that this user is probably looking for a cosmetology hospital. The suggested queries from our model cover this potential information needs and are more likely to be clicked. The second example shows that the clicked document helps our model to understand that the user is interested in LASIK surgery in Hong Kong.

Looking at the attentions paid to queries and words, we can see that in general, those corresponding to the key concepts in the session are captured with more attention. For example, in the second session, the user is likely looking for a place for LASIK surgery, rather than general information explaining bad sight. So Q2 captures less intention than Q1 and Q3. The generated query suggestions reflect this. We can also see that even if any query in the session can obtain some attention weight, in general, the latest query tends to have a higher weight, which is intuitive: the search intent in a session can evolve and the last query better reflects the current intent than an early one.

The above observations show that the attention mechanisms on queries and words can successfully capture the most important elements. However, as we explained earlier, the mechanisms can be fooled by the noise queries in a session, especially when it is long. More investigations are required to detect the true intent of the user. Our model is able to figure out these queries and assign them with lower attention weights. As Q2 in *Context* 2, bad sight is short and vague to infer the information needs. It is not helpful for search intent modeling and thus assigned with a low attention weight by our model.

5   Conclusion

In this paper, we propose a hierarchical attention network (HAN) to explicitly model user search behavior by using not only the issued queries but also the content of clicked documents. HAN encodes queries and clicked documents with two recurrent neural networks and produces a context-aware session embedding hierarchically. The predicted next query is therefore expected to better reflect information needs in the suggestions.

An essential problem of incorporating clicked documents lies in how to select the pivot and informative words. Similarly, identifying strongly relevant queries in a session is another important challenge to infer user information needs. To address these problems, two levels of attention mechanisms are employed to automatically capture the differences without manually selecting pivots. Experiments conducted on a large-scale commercial query log demonstrate the effectiveness of our model. Compared to the model that only uses the issued queries (HRED), our model obtains better performance due to the utilization of click feedback and the attention mechanism.

For future work, we plan to integrate query suggestion to the existing ranking models. By actively rewriting the issued queries with our query suggestion model, the ranking of search results may produce better search results. Another important issue we will investigate is how to better determine the pivot words from the most relevant queries in a noisy session. A more sophisticated topical similarity measure could be integrated in the attention mechanism.

Acknowledgements. This work is supported by Natural Science Foundation of China (Grant No. 61622208, 61732008, 61532011) and National Key Basic Research Program (2015CB358700).

References

1. Gao, B.J., Anastasiu, D.C., Jiang, X.: Utilizing user-input contextual terms for query disambiguation. In: ACL (2010)
2. Sordoni, A., Bengio, Y., Vahabi, H., Lioma, C., Grue Simonsen, J., Nie, J.-Y.: A hierarchical recurrent encoder-decoder for generative context-aware query suggestion. In: CIKM (2015)
3. Song, R., Luo, Z., Nie, J.-Y., Yu, Y., Hon, H.-W.: Identification of ambiguous queries in web search. IPM **45**, 216–229 (2009)
4. Song, Y., Zhou, D., He, L.: Post-ranking query suggestion by diversifying search results. In: SIGIR (2011)
5. He, Q., et al.: Web query recommendation via sequential query prediction. In: ICDE 2009 (2009)
6. Dang, V., Croft, B.W.: Query reformulation using anchor text. In: WSDM (2010)
7. Huang, Z., Cautis, B., Cheng, R., Zheng, Y.: KB-enabled query recommendation for long-tail queries. In: CIKM (2016)
8. Li, L., Deng, H., Dong, A., Chang, Y., Baeza-Yates, R., Zha, H.: Exploring query auto-completion and click logs for contextual-aware web search and query suggestion. In: WWW (2017)
9. Chen, W., Cai, F., Chen, H., de Rijke, M.: Personalized query suggestion diversification. In: SIGIR (2017)
10. Liu, Y., Miao, J., Zhang, M., Ma, S., Liyun, R.: How do users describe their information need: query recommendation based on snippet click model. Expert. Syst. Appl. **38**(11), 13847–13856 (2011)
11. Huang, C.-K., Chien, L.-F., Oyang, Y.-J.: Relevant term suggestion in interactive web search based on contextual information in query session logs. JAIST **54**, 638–649 (2003)
12. Wang, X., Zhai, C.X.: Mining term association patterns from search logs for effective query reformulation. In: CIKM (2008)
13. Liao, Z., Jiang, D., Chen, E., Pei, J., Cao, H., Li, H.: Mining concept sequences from large-scale search logs for context-aware query suggestion. ACM Trans. Intell. Syst. Technol. **3**, 17 (2011)

14. Jiang, J.-Y., Ke, Y.-Y., Chien, P.-Y., Cheng, P.-J.: Learning user reformulation behavior for query auto-completion. In: SIGIR (2014)
15. Diaz, F., White, R., Buscher, G., Liebling, D.: Robust models of mouse movement on dynamic web search results pages. In: CIKM (2013)
16. Liu, Y., Wang, C., Zhou, K., Nie, J., Zhang, M., Ma, S.: From skimming to reading: a two-stage examination model for web search. In: CIKM (2014)
17. Zhou, B., Liu, Y., Zhang, M., Jin, Y., Ma, S.: Incorporating web browsing activities into anchor texts for web search. IR **14**, 290–314 (2011)
18. Sun, A., Lou, C.-H.: Towards context-aware search with right click. In: SIGIR (2014)
19. Joachims, T., Granka, L., Pan, B., Hembrooke, H., Radlinski, F., Gay, G.: Evaluating the accuracy of implicit feedback from clicks and query reformulations in web search. TOIS **25**, 7 (2007)
20. Chapelle, O., Zhang, Y.: A dynamic Bayesian network click model for web search ranking. In: WWW, pp. 1–10 (2009)
21. Chung, J., Gulcehre, C., Cho, K.H., Bengio, Y.: Empirical evaluation of gated recurrent neural networks on sequence modeling. Eprint arxiv (2014)
22. Rocktäschel, T., Grefenstette, E., Hermann, K.M., Kočiský, T., Blunsom, P.: Reasoning about entailment with neural attention. In: ICLR (2015)
23. Wu, Q., Burges, C.J.C., Svore, K.M., Gao, J.: Adapting boosting for information retrieval measures. Inf. Retr. **13**, 254–270 (2010)
24. Cho, K., et al.: Learning phrase representations using RNN encoder-decoder for statistical machine translation. In: EMNLP (2014)

Short Papers from AIRS 2017

Assigning NDLSH Headings to People on the Web

Masayuki Shimokura$^{(\boxtimes)}$ and Harumi Murakami

Graduate School for Creative Cities, Osaka City University,
3-3-138, Sugimoto, Sumiyoshi, Osaka 558-8585, Japan
shimokura@gmail.com

Abstract. We investigate a method that assigns National Diet Library Subject Headings (NDLSH) to the results of web people searches to help users select and understand people on the web. NDLSH is a controlled subject vocabulary list compiled and maintained by the National Diet Library (NDL) as a subject access tool. By assigning NDLSH headings to people, well-formed keywords can be assigned, and exploratory searches using related terms are possible. We examined the following combination of factors: (a) web-page rank (the number of pages), (b) position inside the HTML, (c) synonyms, and (d) document frequency. We report our experimental results for 405 combination patterns ($5 \times 9 \times 3 \times 3$) using our 80-person dataset. Overall, under our experimental settings, the best combination was (a) the top ten pages, (b) 100 characters before and after a person's name (i.e., 200 characters), (c) half weight for synonyms, and (d) document frequency divided by number of web pages.

Keywords: Web people search · Subject headings assignment
NDLSH · Experiment

1 Introduction

The popularity of web people searches continues to rise as the number of people increases about whom the web can provide information. If the list of web people search results is merely "person 1, person 2, and so on," users have difficulty determining which person they should select. Appropriate labels shown with people should help users select the person they want.

In our previous work [1], we assigned people location information [2], vocation-related information [3], and library classification numbers [4]. In this paper, we investigate a method that assigns National Diet Library Subject Headings (NDLSH) to the results of web people searches to help users select and understand people on the web. The NDLSH [5] is a controlled subject vocabulary list compiled and maintained by the National Diet Library (NDL) as a subject access tool. By assigning NDLSH headings to people, well-formed keywords can be assigned, and exploratory searches using related terms are possible.

© Springer Nature Switzerland AG 2018
Y.-H. Tseng et al. (Eds.): AIRS 2018, LNCS 11292, pp. 189–195, 2018.
https://doi.org/10.1007/978-3-030-03520-4_18

We examined the following combination of factors: (a) web-page rank (the number of pages), (b) position inside the HTML, (c) synonyms, and (d) document frequency. We report our experimental results for 405 combination patterns ($5 \times 9 \times 3 \times 3$) using our 80-person dataset.

Below, we explain our experiment in Sects. 2 and 3, and the significance of our research in Sect. 4.

2   Method

2.1   Procedure

As queries, we used 20 Japanese names from a related work [6] and obtained 50 web pages per each query via Google Custom Search API. We manually classified these pages into different people and identified 80 separate people. NDLSH headings were assigned to HTML files for each person.

First, we extract headings with variants (synonyms) and delete those with two or fewer single-byte alphanumeric characters, those with only one double-byte character, and those contain "--(two hyphens)" because they are less important terms and/or not very useful for text-matching.

With regard to terms (headings and synonyms), since longer strings provide more specific meaning, we count the number of terms (headings and synonyms) from the longer ones in the HTML documents without tags in the following conditions: (a) and (b). For example, when processing the "artificial intelligence" character string in a document, the term "artificial intelligence" is counted but not the word "intelligence." After counting the terms (headings and synonyms), the scores of the headings are calculated.

We prepared the following four types of combination conditions:

(a) Search ranking of web pages: five patterns of top 1, 3, 5, 10, and all pages for each person.
(b) Position in HTML documents: nine patterns of title only, full text, and the 20, 40, 60, 80, 100, 150, or 200 characters before and after a person's name.
(c) Synonyms: three patterns that don't use synonyms, using synonyms with identical weights as the headings, and using synonyms with half of the weight of the headings.
(d) Document frequency of headings and synonyms: three patterns of doing nothing, multiplying document frequency (df)/total number of used documents (N) (i.e., multiplying df/N), and multiplying total number of used documents (N)/document frequency (df) (i.e., multiplying N/df). The last one is rephrased as multiplying inverse document frequency (i.e., multiplying idf).

When these conditions are combined, they become $5 \times 9 \times 3 \times 3 = 405$ patterns. Figure 1 shows an example of the score calculation for headings. A heading having with the highest score is assigned to the corresponding person. If no heading is assigned, the answer becomes "none".

Score of NDLSH heading "SH1" for HTML documents of person A

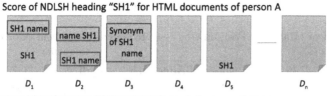

Pattern (a) Top five pages, (b) Characters before and after a person's name,
(c) Synonyms×0.5, (d) df/N

SH1: 3 times; synonyms of SH1 : 1×0.5 = 0.5

(3+0.5)×3/5 = 3.5×0.6 = 2.1

Pattern (a) Top three pages, (b) Full text, (c) No synonym, (d) N/df

SH1: 4 times

4×3/2 = 4×1.5 = 6.0

Fig. 1. Example of score calculation

2.2 Evaluation

We manually selected the most appropriate NDLSH heading for each person
(79 out of 80 people). For example, a "baseball" heading was selected for Sug-
uru Egawa, a former professional baseball player, and "social psychology" was
selected for Asako Miura, a Kwansei Gakuin University professor.

The following are the evaluation measures:

$$\text{Correctness} = \frac{\text{number of correct NDLSHs assigned automatically}}{\text{number of people}}$$

$$\text{Precision} = \frac{\text{number of correct NDLSHs assigned automatically}}{\text{number of people to whom an NDLSH was assigned automatically}}$$

$$\text{Recall} = \frac{\text{number of correct NDLSHs assigned automatically}}{\text{number of people to whom an NDLSH was assigned manually}}$$

$$\text{F-measure} = \frac{2 \times \text{Precision} \times \text{Recall}}{\text{Precision} + \text{Recall}}$$

$$\text{Accuracy} = \frac{\text{number of correct answers}}{\text{number of people}}$$

When calculating the Accuracy, none is judged correct when there is no
correct NDLSH heading for a person.

3   Results and Analysis

Table 1 shows which pattern had the best correctness. The best pattern for
the whole (80 people) was "top ten pages, 100 characters before and after the
person's name (total 200 characters), 0.5 times for synonyms, and df/N," and
its correctness was 26.3% (21/80).

Table 1. Patterns of high correctness

Number of documents	Ranking	Position	Synonym	Document frequency	Correctness
All	10	Before and After 100 Characters	0.5	df/N	0.263 (21/80)
1	1	Full Text	None	1	0.286 (10/35)
2	3	Before and After 200 Characters	0.5	1	0.333 (4/12)
3 or more	10	Before and After 100 Characters	1	df/N	0.364 (12/33)
11 or more	10	Before and After 100 Characters	1	df/N	0.500 (9/18)
3 to 10	5	Before and After 60 Characters	0.5	df/N	0.267 (4/15)

Since we observed that the trend may differ depending on the number of documents for each person, we classified the number of documents into 1, 2, 3 or more, 11 or more, 3 to 10, and conducted our evaluation. For 35 people with only one document, "full text (not using synonyms)" was best. For 33 people with 3 or more documents, the result was almost the same as for all the people (the difference is only the magnification of synonyms), and the result was the same for 18 people with 11 documents or more.

Table 2. Evaluation for patterns of high correctness

Number of documents	Precision	Recall	F-measure	Accuracy
All	0.276	0.266	0.271	0.263
1	0.286	0.294	0.290	0.286
2	0.333	0.333	0.333	0.333
3 or more	0.364	0.364	0.364	0.364
11 or more	0.500	0.500	0.500	0.500
3 to 10	0.182	0.267	0.216	0.267

Evaluations for high correctness patterns are shown in Table 2. Precision and recall are less than 30% for both all people and people with one document; they are 50% for people with 11 or more documents.

For comparison, we implemented a method using cosine as a baseline. MeCab [7], a Japanese morphological analyzer, was used to extract terms from documents (web pages), headings and synonyms; nouns that contain two or more

Table 3. Evaluation using cosine

Number of documents	Precision	Recall	F-measure	Accuracy
All	0.026	0.025	0.026	0.025

characters were extracted as terms. To match the conditions to our best pattern, we used the top ten pages, 100 characters before and after a person's name (200 characters), synonyms, and document frequency. The weights of the terms extracted from the headings were normalized by frequency $\times$ df/N, and the weights of the terms extracted from the synonyms were 0.5 times normalized frequency $\times$ df/N. Table 3 shows the evaluation results with cosine. Our best pattern significantly outperformed cosine.

Based on the above results, we found that we must narrow down to the top ten rather than all of the whole web pages, use the character strings before and after a person's name rather than the full text of the HTML documents, and weight the words that appear in many documents using synonyms. However, the best approach was counting the number of headings from the full text for those with only one document. These performances were much better than the method using cosine.

4   Related Work and Discussion

We roughly divided the methods for assigning controlled terms to documents into with and without machine learning. Our research falls in the latter classification. Cosine is one of the most commonly used baselines for non-machine learning methods. The results obtained in this research are significantly higher than the baseline.

In research that assigns Nippon Decimal Classification (NDC) numbers to people on the web to develop a person directory [4], we considered two conditions of documents $\times$ methods. The title (texts inside title tags) was best among the document conditions, and the result differs from this research. We believe that this is due to the difference of research purposes and dataset. [8] suggested that the best patterns were different based on the number of documents per person (i.e., 1, 2, 3 or more documents), and this is related to this research.

There is research that assigns labels to people except for our previous work. Wan et al. [9] assigned titles (including vocations) and Mori et al. [10] assigned keywords to person clusters. WePS-2/3 [11] conducted competitive evaluation on person attribute extraction on web pages. No such research has assigned subject headings to people on the web.

Our experimental results revealed the best pattern for assigning NDLSH headings to people on the web. To the best of our knowledge, this is the first research that assigns library subject headings to people on the web.

5 Conclusions

We investigated a method that assigns NDLSH headings to the results of web people searches to help users select and understand people on the web. We examined the following combination of factors: (a) web-page rank, (b) position inside HTML, (c) synonyms, and (d) document frequency. We reported the results of our experiment for 405 patterns with an 80-person dataset. Overall, under our experimental settings, the best combination was (a) the top ten pages, (b) 100 characters before and after a person's name (i.e., 200 characters), (c) half weight for synonyms, and (d) document frequency divided by the number of web pages.

Future work will examine the use of stop lists and such terms as broader, narrower, and related terms. We also need to evaluate our method for other headings (except for highest score) and with large and diverse datasets.

Acknowledgements. This work was supported by JSPS KAKENHI Grant Number 25330385, 16K00440.

References

1. Murakami, H., Ueda, H., Kataoka, S., Takamori, Y., Tatsumi, S.: Summarizing and visualizing web people search results. In: Proceedings of the Second International Conference on Agents and Artificial Intelligence (ICAART 2010), vol. 1, pp. 640–643. INSTICC Press (2010)
2. Murakami, H., Takamori, Y., Ueda, H., Tatsumi, S.: Assigning location information to display individuals on a map for web people search results. In: Lee, G.G., et al. (eds.) AIRS 2009. LNCS, vol. 5839, pp. 26–37. Springer, Heidelberg (2009). https://doi.org/10.1007/978-3-642-04769-5_3
3. Ueda, H., Murakami, H., Tatsumi, S.: Assigning vocation-related information to person clusters for web people search results. In: Proceedings of the 2009 Global Congress on Intelligent Systems (GCIS 2009), vol. 4, pp. 248–253. IEEE Press, New York (2009)
4. Murakami, H., Ura, Y., Kataoka, Y.: Assigning library classification numbers to people on the web. In: Banchs, R.E., Silvestri, F., Liu, T.-Y., Zhang, M., Gao, S., Lang, J. (eds.) AIRS 2013. LNCS, vol. 8281, pp. 464–475. Springer, Heidelberg (2013). https://doi.org/10.1007/978-3-642-45068-6_40
5. Cataloging Tools and Resources. http://www.ndl.go.jp/en/data/classification_subject.html
6. Sato, S., Kazama, K., Fukuda, K., Murakami, K.: Distinguishing between people on the web with the same first and last name by real-world oriented web mining. IPSJ Trans. Databases **46**(8), 26–36 (2005)
7. MeCab: Yet Another Part-of-Speech and Morphological Analyzer. http://taku910.github.io/mecab/
8. Murakami, H., Ura, Y., Kataoka, Y.: Assigning library classification numbers to people on the web and developing people-search directory. Trans. Inst. Syst. Control Inf. Eng. **29**(2), 51–64 (2016). in Japanese
9. Wan, X., Gao, J., Li, M., Ding, B.: Person resolution in person search results: WebHawk. In: Proceedings of the Fourteenth ACM Conference on Information and Knowledge Management (CIKM 2005), pp. 163–170. ACM Press, New York (2005)

10. Mori, J., Matsuo, Y., Ishizuka, M.: Personal keyword extraction from the web. J. Jpn. Soc. Artif. Intell. **20**, 337–345 (2005)
11. Artiles, J., Borthwick, A., Gonzalo, J., Sekine, S., Amigo, E.: WePS-3 evaluation campaign: overview of the web people search clustering and attribute extraction tasks. In: CLEF 2010 (2010)

MKDS: A Medical Knowledge Discovery System Learned from Electronic Medical Records (Demonstration)

Hen-Hsen Huang[1(✉)], An-Zi Yen[1], and Hsin-Hsi Chen[1,2]

[1] Department of Computer Science and Information Engineering, National Taiwan University, Taipei, Taiwan
{hhhuang, azyen}@nlg.csie.ntu.edu.tw,
hhchen@ntu.edu.tw
[2] MOST Joint Research Center for AI Technology and All Vista Healthcare, National Taiwan University, Taipei, Taiwan

Abstract. This paper presents a medical knowledge discovery system (MKDS) that learns the medical knowledge from electronic medical records (EMRs). The distributed word representations model the relations among medical concepts such as diseases and medicines. Four tasks, including spell checking, clinical trait extraction, analogical reasoning, and computer-aided diagnosis, are demonstrated in our system.

Keywords: Medical knowledge discovery · Medical records
Distributed word representation

1 Introduction

Clinical decision supporting (CDS) systems provide physicians with professional knowledge for clinical decision-making [3]. The popular CDS systems such as UpToDate and Micromedex are online database systems containing the information of drugs, diseases, diagnosis, symptoms, exams, surgeries, and so on. After a user enters a keyword, e.g., "leukemia", the prognosis, the symptoms, the exams, and the treatments for each subtype of leukemia are shown. Such information is very useful for physicians and students to do clinical case studies.

On major CDS systems, the contents are human edited. Domain experts organize medical information into a database based on their prior knowledge. In this paper, we show an approach to discover the relations among medical concepts from medical documents, and apply the medical knowledge to aid some medical applications.

Electronic medical records (EMRs) written by physicians is a rich source of professional knowledge [3]. This work presents a medical knowledge discovery system (MKDS)[1], which aids to discover the relationships among medical concepts from EMRs. Different from the approaches based on traditional information retrieval [9], our method utilizes the skip-gram model, which has been shown to represent lexical term

---

[1] http://nlg18.csie.ntu.edu.tw:8181.

© Springer Nature Switzerland AG 2018
Y.-H. Tseng et al. (Eds.): AIRS 2018, LNCS 11292, pp. 196–202, 2018.
https://doi.org/10.1007/978-3-030-03520-4_19

semantics effectively [8], to capture the relatedness among medical concepts. We demonstrate the uses of word vectors in four tasks in our system, including spell checking, clinical trait extraction, analogical reasoning, and computer-aided diagnosis. Compared to traditional approaches, our method addresses healthcare issues like literature analysis, prognosis, and patent management.

The main contributions of this work are three folds: (1) We show how the word vectors learned from EMRs can help capture medical relationships among medicines, surgeries, diagnosis, and exams. (2) Four applications based on the extracted knowledge are demonstrated with an instructive and educational system. (3) We release word vectors trained from medical documents[2]. That can be applied to various clinical NLP applications.

2 Related Work

Knowledge discovery in medical documents is an attractive topic. The medical tracks in TREC 2011 and 2012 deal with the task of information retrieval on EMRs [16]. In i2b2 NLP challenges, the shared tasks, including medical term extraction and classification of relations between medical concepts, are explored with EMRs [12]. EMRs are used in many applications including assertion classification [6], clinical trait extraction [2], co-reference resolution [14], temporal analysis [13], pneumonia identification [1], phenotyping [4, 10, 11], and outpatient department recommendation [5].

Neural network models such as word2vec [8] represent a word as a vector in a low-dimensional space, where semantic relatedness can be measured. In addition, the property of linguistic regularity is also observed [7, 9]. Using distributed word representation is also shown to improve the clinical concept extraction [15].

3 Resources

The experimental dataset is composed of the EMRs from National Taiwan University Hospital (NTUH). As shown in Table 1, total 113,625 medical records are collected in five years. An EMR contains three main sections, the chief complaint, the brief history, and the course and treatment. The chief complaint denotes the purpose of the patient's

Table 1. Statistics of the EMRs in the NTUH corpus.

Department	# Records	Department	# Records
Dental	1,253	Ophthalmology	3,400
Internal Medicine	34,396	Obstetrics & Gynecology	5,679
Oncology	4,226	Dermatology	1,258
Pediatrics	11,468	Ear, Nose & Throat	7,680
Surgery	23,303	Rehabilitation	1,935
Urology	5,818	Orthopedics	8,814
Neurology	2,739	Psychiatry	1,656

---

[2] http://nlg18.csie.ntu.edu.tw/mkds/medical.w2v.

visit, e.g., "headache for a week". The brief history presents the background information of the patient like her/his age, gender, and past diseases. The course and treatment note the treatment such as medicines, surgeries, and exams for the patient.

The Unified Medical Language System (UMLS) is adopted as the ontology of medical terms. Five types of medical terms including drug, exam, surgery, diagnosis, and body are collected from the UMLS.

4 Medical Term Representation

We learn the distributed word representations from the preprocessed EMRs.

4.1 Preprocessing

EMRs are usually written in a rush, thus noise is unavoidable. Spelling errors, grammatical errors, and abbreviations are common in EMRs. Preprocessing is performed to deal with the related issues.

Age. Six age patterns are found in EMRs: "#-year-old", "#-years-old", "# year old", "# years old", "# y/o", and "#-y-o". All the ages are standardized to four groups: under 15 years old, 16 to 45 years old, 46–60 years old, and beyond 60 years old.

Gender. Eight gender patterns are found in the NTUH corpus: woman, lady, female, girl, man, gentleman, male, and boy. We standardize them to two types of gender.

Medical Terms. In EMRs, the UMLS terms are identified as medical terms. The phrases and the compounds such as "coronary artery disease" are treated as single terms. For each of major diseases such as "diabetes" or "coronary artery disease", we merge all its alias and abbreviations into one. For each drug, we replace all its trade names with the generic name, and discard the dosage form.

4.2 Representation Learning

The Skip-gram model is used to train the word representations of medical terms. The size of dimension is 500, the context window is 10, and negative sampling is used with size 5. In this way, a medical term t is represented by a vector v with length 500, and the relatedness between two medical terms t_1 and t_2 can be measured by the cosine similarity of their vectors v_1 and v_2.

5 Medical Knowledge Discovery

The learned medical word vectors are employed to four tasks in our system.

5.1 Spell Checking

We collect the frequent OOV terms from the NTUH corpus. We check if an OOV term t has spelling error, and correct it by using the medical word vectors. First, we fetch 10 medical terms most related to *t*. The longest common subsequence (LCS) algorithm is performed to measure the overlap between a candidate *c* and *t*. The most related candidate *c* with an overlap of at least 90% is chosen as the counterpart of *t*. Table 2 lists the top 5 spelling errors. The homophone error "leukovorin" is corrected. In EMRs, the exam colonoscopy is often misspelled as colonscope, which is the instrument used in the exam. Full list is available on the website of MKDS.

Table 2. Most frequent spelling errors in the NTUH corpus.

Error	Correction	# Occurrences
Leukovorin	Leucovorin	6,013
Lower extremity	Lower extremities	2,308
Colonscope	Colonscopy	1,801
Esophageal varix	Esophageal varices	1,709
Hypermetabolism	Hypermetabolic	1,391

5.2 Clinical Trait Extraction

MKDS extracts the most related medical terms in each type for an input term. To evaluate the performance, domain knowledge is consulted by the national cancer institute website (NCI). Six common cancer types, including lung cancer, female breast cancer, colon cancer, prostate cancer, esophageal cancer, and leukemia, are chosen. From the NCI data, the drugs approved to treat each type of cancers are listed. These medications form the ground truth for evaluation. Because not all the drugs listed on NCI are used in NTUH, the intersections of NTUH and NCI are computed.

The performance is measured by average precision at 10 (ap@10). Table 3 shows the drugs suggested by our system for each type of cancers. Different from the NCI list, our clinical traits are weighted. Among 63 drugs for breast cancer, the top one, tamoxifen citrate, is the most common hormone therapy of the estrogen receptor positive (ER+) breast cancer. For lung cancer, gefitinib is a kind of targeted therapy for non-small cell lung cancers with mutated epidermal growth factor receptor (EGFR). Gefitinib is a common drug for lung cancer treatment in NTUH because EGFR mutations are much more prevalent in Asia.

Table 3. Performance of clinical trait extraction.

Cancer	ap@10	Approved drugs for treatment
Lung	64.15%	Gefitinib, erlotinib, pemetrexed disodium heptahydrate
Breast	42.19%	Tamoxifen citrate, zoladex, halcion, simethicone, goserelin
Colon	53.14%	Oxaliplatin, cetuximab, capecitabine, bevacizumab
Prostate	43.19%	Flutamide, bicalutamide, deflux
Esophageal	12.86%	Docetaxel
Leukemia	39.88%	Imatinib mesylate, idarubicin, daunorubicin, cytarabine

5.3   Analogical Reasoning

Analogical reasoning based on word vectors can be simplistically accomplished by vector arithmetic like 3COSADD [9]. Our system answers the analogy questions like $t_1:t_2 \approx t_3:t_4$, where t_4 is unknown. For example, "tamoxifen citrate" is one of answers to the question "lung cancer:gefitinib $\approx$ breast cancer:?".

Referred to Table 3, we compose questions $t_1:t_2 \approx t_3$ with the top one medication for each of the six cancer types. Table 4 shows some promising analogies. The relation between a cancer type and its drug can be captured by vector offsets.

Table 4. Analogical reasoning between cancers and medications.

t_1 (cancer)	t_2 (drug)	t_3 (cancer)	t_4 (drug)
Prostate	Flutamide	Lung	Gefitinib
Prostate	Flutamide	Breast	Tamoxifen citrate
Leukemia	Imatinib mesylate	Colon	Oxaliplatin
Lung	Gefitinib	Leukemia	Imatinib mesylate
Breast	Tamoxifen citrate	Lung	Gefitinib

5.4   Computer-Aided Diagnosis

In this task, a user inputs a chief complaint, and the system responses the most related diagnosis and treatment. Rather than a single medical term, a chief complaint is a short sentence or a statement, where medical and non-medical terms are intermixed. Figure 1 shows a snapshot of the computer-aided diagnosis in MKDS.

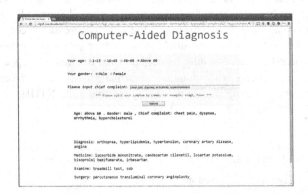

Fig. 1. Snapshot of the use of computer-aided diagnosis.

6 Conclusions

This work demonstrates four applications of distributed medical term representations. The properties of the skip-gram model such as semantic relatedness and linguistic regularity are utilized to implement the four tasks. Compared to the traditional pre-trained word vectors, our approach acquires the knowledge from the real-world medical records. With more and newer EMRs, our system will be more instructive.

Acknowledgements. This research was partially supported by Ministry of Science and Technology, Taiwan, under grants MOST 106-3114-E-009-008 and MOST-105-2221-E-002-154-MY3, and National Taiwan University under grant NTUCCP-106R891305.

References

1. Bejan, C.A., Vanderwende, L., Wurfel, M.M., Yetisgen-Yildiz, M.: Assessing pneumonia identification from time-ordered narrative reports. In: Proceedings of 2012 AMIA Annual Symposium, pp. 1119–1128 (2012)
2. Davis, M.F., Sriram, S., Bush, W.S., Denny, J.C., Haines, J.L.: Automated extraction of clinical traits of multiple sclerosis in electronic medical records. J. Am. Med. Inform. Assoc. **20**(2), 334–340 (2013)
3. Demner-Fushman, D., Chapman, W.W., McDonald, C.J.: What can natural language processing do for clinical decision support? J. Biomed. Inform. **42**(5), 760–772 (2009)
4. Hripcsak, G., Albers, D.J.: Next-generation phenotyping of electronic health records. J. Am. Med. Inform. Assoc. **20**(1), 117–121 (2013)
5. Huang, H.-H., Lee, C.-C., Chen, H.-H.: Mining professional knowledge from medical records. In: Ślęzak, D., Tan, A.-H., Peters, James F., Schwabe, L. (eds.) BIH 2014. LNCS (LNAI), vol. 8609, pp. 152–163. Springer, Cham (2014). https://doi.org/10.1007/978-3-319-09891-3_15
6. Kim, Y., Riloff, E., Meystre, S.M.: Improving classification of medical assertions in clinical notes. In: Proceedings of the 49th Annual Meeting of the Association for Computational Linguistics (ACL): Short Papers, pp. 311–316 (2011)
7. Levy, O., Goldberg, Y.: Linguistic regularities in sparse and explicit word representations. In: Proceedings of the 18th Conference on Computational Language Learning, pp. 171–180 (2014)
8. Mikolov, T., Chen, K., Corrado, G., Dean, J.: Efficient estimation of word representations in vector space. In: ICLR Workshop Papers (2013)
9. Mikolov, T., Yih, W.T., Zweig, G.: Linguistic regularities in continuous space word representations. In: Proceedings of NAACL-HLT, pp. 746–751 (2013)
10. Pathak, J., Kho, A.N., Denny, J.C.: Electronic health records-driven phenotyping: challenges, recent advances, and perspectives. J. Am. Med. Inform. Assoc. **20**(e2), e206–e211 (2013)
11. Shivade, C., et al.: A review of approaches to identifying patient phenotype cohorts using electronic health records. J. Am. Med. Inform. Assoc. **21**, 221–230 (2014)
12. Stubbs, A., Kotfila, C., Xu, H., Uzuner, Ö.: Identifying risk factors for heart disease over time: overview of 2014 i2b2/UTHealth shared task track 2. J. Biomed. Inform. **58**, S67–S77 (2015)

13. Sun, W., Rumshisky, A., Uzuner, O.: Evaluating temporal relations in clinical text: 2012 i2b2 challenge. J. Am. Med. Inform. Assoc. **20**(5), 806–813 (2013)
14. Uzuner, O., Bodnari, A., Shen, S., Forbush, T., Pestian, J., South, B.R.: Evaluating the state of the art in coreference resolution for electronic medical records. J. Am. Med. Inform. Assoc. **19**(5), 786–791 (2012)
15. De Vine, L., Kholghi, M., Zuccon, G., Sitbon, L., Nguyen, A.: Analysis of word embeddings and sequence features for clinical information extraction. In: Proceedings of the 13th Annual Workshop of the Australasian Language Technology Association (2015)
16. Voorhees, E.M., Hersh, W.: Overview of the TREC 2012 medical records track. In: Proceedings of the 21st Text REtrieval Conference (2012)

Predicting Next Visited Country of Twitter Users

Muhammad Syafiq Mohd Pozi[1]([⊠]), Yuanyuan Wang[2], Panote Siriaraya[1],
Yukiko Kawai[1], and Adam Jatowt[3]

[1] Kyoto Sangyo University, Motoyama, Kamigamo, Kita-ku, Kyoto 603-8555, Japan
{syafiq,kawai}@cc.kyoto-su.ac.jp, spanote@gmail.com
[2] Yamaguchi University, 2-16-1 Tokiwadai, Ube, Yamaguchi 755-8611, Japan
y.wang@yamaguchi-u.ac.jp
[3] Kyoto University, Yoshida-homachi, Sakyo-ku, Kyoto 606-8501, Japan
adam@dl.kuis.kyoto-u.ac.jp

Abstract. We develop a classification model to predict which country
will be visited next by Twitter users. In our model we incorporate a range
of spatial and temporal attributes as well as we use language as one of
additional, novel attributes for predicting user movement. We found that
these attributes can be used to obtain a consistent classification model.

Keywords: Mobility analysis · Classification · Microblogging

1 Introduction

One trending research theme in social network analysis is how to model human
mobility behavior [7]. This line of research can be divided to either model the
trajectory of an individual user [2] or group of users [6]. The model is usually
derived from the social network metadata such as geotagged messages, user
followers, demographic and spatiotemporal data [5].

In this paper, we are modeling user traveling behavior in terms of visited
countries based on the classification approach. There are two possible settings
in this research, the first one is to model the characteristics of travelers that is
to distinguish travelers from stationary users. The second one is to predict the
trajectory of travelers [1]. In our research, we assume that each user in our data
is a traveler, hence, we choose the second setting.

Our contribution can be considered as the novel setting of the classification
task: predicting the country of the next travel and the prediction model designed
by considering spatial and temporal aspects along with user languages as novel
features added for improving prediction accuracy.

2 Methodology

In this section, we describe the methodology used in our research. The high level
abstraction of the prediction model is illustrated in Fig. 1.

© Springer Nature Switzerland AG 2018
Y.-H. Tseng et al. (Eds.): AIRS 2018, LNCS 11292, pp. 203–209, 2018.
https://doi.org/10.1007/978-3-030-03520-4_20

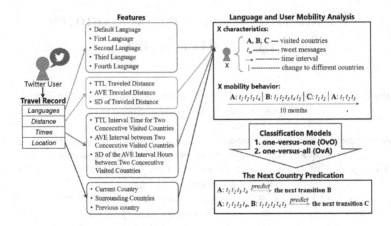

Fig. 1. Language and spatio-temporal analysis of Twitter data.

2.1 Data Collection and Preprocessing

Approximately 10 months of geo-tagged Twitter data was gathered from 1st May 2016 to 7th March 2017, consists of 2.16×10^7 geotagged tweets produced by 1.24×10^6 unique Twitter's users in southwest Europe. The reason we decided to focus on the subarea of Europe, is because this continent is characterized by the large density of diverse languages, and, thanks to EU and to the Schengen agreements, by free, unrestricted travel between the EU member countries. 24 languages spoken in Europe are official languages.

2.2 Feature Extraction

Each record in the database that stores the above-described collected data corresponds to a particular tweet and tweets are sorted from the earliest to the latest date (Sec. 3.3 describes the details of the data model). We divide the classification features to be used into the following groups: language (1–5), distance (6–8), time (9–11) and location (12–14). The features and the class to be predicted (the next country indicated by number 15) are as follows:

1. **Default language:** The user default language defined in user profile in her Twitter account.
2. **First language:** The main language of a user. The main language is defined as the language of the largest number of tweets issued by a user (e.g., a user tweeting in French and Spanish with 1000 and 500 tweets, respectively, is considered to have Spanish as his/her first language).
3. **Second language:** The second most common language of a user. The second language is defined as the language of the second largest number of tweets by the user.
4. **Third language:** The third language of the user.
5. **Fourth language:** The fourth language of the user.

Motivation (1–5). Language is highly associated with a particular country, i.e. most European countries have different official languages. Hence, language should be suggestive of a country users will be visiting next (e.g., when returning home from a travel abroad, or when communicating with foreign friends who live abroad before visiting them). In general, we believe that the capability of user to converse in particular languages should affect their trajectory model.

6. **Total traveled distance:** The total traveled distance of users, where each distance is computed on the basis of locations of consecutive tweets.
7. **Average traveled distance:** Average traveled distance d is computed based on the locations of the consecutive tweets of the same user. Then, the average, $\bar{x}$, is calculated based on all the N distances derived from tweet stream of the user: $\bar{x} = (\sum_{i=1}^{N} d_i)/N$.
8. **Standard deviation of the traveled distance:** The standard deviation, σ of the distance is computed based on all N distances: $\sigma = \sqrt{\frac{1}{N-1} \sum_{i=1}^{N} (d_i - \bar{x})^2}$.

Motivation (6–8). The traveled distance for each user can also be used to identify the user capability of traveling to many countries, especially, to far away countries. As the users' traveled distance varies over time, they may be visiting the near or far away countries.

9. **Total interval time for consecutively visited countries:** The total traveled times in hours of each user, where each travel time is represented by the time between two consecutive tweets generated from two different countries.
10. **Average interval between consecutively visited countries (expressed in hours):** it is computed based on the location of two consecutive tweets from two consecutive different countries. Then, the average is computed based on all N of interval hours: $\bar{x} = (\sum_{i=1}^{N} h_i)/N$.
11. **Standard deviation of the average interval hours between two consecutively visited countries:** it is computed based on the location of two consecutive tweets from two consecutive different countries. Then, the standard deviation is also computed based on all N of interval hours: $\sigma = \sqrt{\frac{1}{N-1} \sum_{i=1}^{N} (h_i - \bar{x})^2}$.

Motivation (9–11). The traveled time interval for each user can be used to identify the user capability of traveling to many countries. A user could stay in a country for a short term, possibly, due to the transition between flights, or for a long term due to business trip or vacation.

12. **Current country:** The country from where each individual tweet has been generated. The countries are listed in Table 1.
13. **Surrounding countries:** Top 3 nearest countries of the current country. The list of surrounding countries are obtained from *www.gomapper.com* and shown in Table 1. Each surrounding country is additionally defined by: (a) the population of the country and (b) the official language of the country.

14. **Previous country:** Previously visited country prior to the current country.
15. **Next country:** Next country to be visited after the current country. This defines the class of the classification model.

Table 1. Surrounding countries of the current country.

Countries	Surrounding Countries
Austria	Czech Republic, Croatia, Germany
Belgium	Netherlands, United Kingdom, Germany
Switzerland	France, Austria, Belgium
Czech Republic	Austria, Germany, Croatia
Germany	Netherlands, Czech Republic, Belgium
Denmark	Netherlands, German, Belgium
Spain	Portugal, France, Switzerland
France	Czech Republic, Belgium, United Kingdom
United Kingdom	Belgium, Netherlands, Ireland
Croatia	Austria, Italy, Czech Republic
Ireland	United Kingdom, Netherlands, Switzerland
Italy	Croatia, Austria, Switzerland
Netherlands	Belgium, Germany, Denmark
Portugal	Spain, France, Ireland

Motivation (12–14). Users traveling behavior can be affected by which country they have visited before, and the current country they are in. The surrounding countries are likely to be visited next and the larger the population of those countries, the higher could be the chance of the visit. Also, if languages spoken in those countries are same as ones of users the possibility of the visit could increase.

2.3 Data Cleaning and Modeling

Once we obtained the data, as explained in Sect. 2.2, we first removed noisy or useless records based on the following filters (Table 2):

1. **User total tweets:** Users that have less than 10 tweets or more than 10,000 tweets for the entire period of study are removed from the dataset. These users are assumed either to be inactive users or bots.
2. **Number of visited countries:** In this research, we are only interested in users that are traveling to at least more than two countries, that is, users who have at least two tweets originating from two different countries. Tweets that belong to users that were tweeting only from the same country are removed.

Table 2. The resulting data after the data cleaning process.

Description	Total tweets	Total users
Original dataset	26,137,207	1,232,122
Remove users that tweet less than 10 or more than 10,000 times	15,387,205	274,934
Remove users that have 0 km total travel distance	6,753,506	122,929

Data Modeling. Based on the obtained data we define a travel record. A travel record consists of a tweet set of a user from the start date of the dataset until the first tweet issued from the different country when compared to the user's current country (i.e., the point of country transition). Note that a single user could have several travel records based on several visited countries. Based on the requirement for users to change countries during the timespan of our dataset we obtained in total 92,195 users with 270,424 travel records.

3   Experimentation and Analysis

As the processed data is relatively large, in order to make the training time feasible, random forest [4] was selected for this experiment due to its fast computation. Furthermore, as this is a multiclass classification task, we compared random forest classification performance based on two learning approaches:

1. One-versus-One (OvO): The dataset is partitioned into a set of $K(K-1)/2$ class respectively. Then, the classifier is trained against each set, hence obtaining $K(K-1)/2$ classification model. Finally, all these models are used on unseen data for prediction, outputting the most probable class value based on the classification model. The most frequent class value is voted as the final class for that data.
2. One-versus-All (OvA): Like OvO, the set however consists of a particular class against other classes, hence obtaining K classification models. Finally, all these models are used on unseen data for prediction, outputting the most probable class value. The most frequent class value is voted as the final class for that data.

The classifier implementation is based on WEKA version 3.9 [3]. The default parameters of both the learning algorithms were used, in which the maximum depth and features of each tree is set to unlimited, and total number of training iterations is set to 100. Total classification accuracy and weighted F-measure are employed to compare the classification performance. The evaluation is obtained by 10-fold stratified cross-validation on the partitions of the whole test set.

Table 3 displays the results of OvO and OvA random forest classifier for each class. GB, ES, and FR are the top accurately classified next countries while AT, HR and CZ are the top falsely classified next countries. This is probably due to the class distribution as shown in Table 3 being significantly high for GB,

Table 3. The true positive rate in % for each class based on one-versus-one (OvO) and one-versus-all (OvA) learning approaches.

Class	#Travel records	OvO	OvA
Austria (AT)	4,769	16.90	18.30
Belgium (BE)	14,907	42.30	42.80
Switzerland (CH)	10,512	25.20	26.60
Czech Republic (CZ)	4,890	20.30	23.40
Germany (DE)	27,984	46.80	46.80
Denmark (DK)	3,747	20.30	24.20
Spain (ES)	39,268	65.70	65.60
France (FR)	38,912	52.80	51.40
United Kingdom (GB)	58,231	71.70	71.10
Croatia (HR)	2,179	6.80	10.10
Ireland (IE)	8,250	42.90	45.70
Italy (IT)	25,005	45.00	45.20
Netherlands (NL)	22,166	46.50	46.90
Portugal (PT)	9,604	47.90	48.90
Overall accuracy	**270,424**	**52.70**	**52.80**

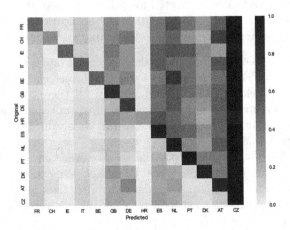

Fig. 2. Normalized confusion matrix based on one-versus-all random forest classifier.

ES and FR but significantly lower for AT, HR and CZ. As expected countries with large populations (DE, GB) and/or culturally grounded higher rates of geo-located Tweets (e.g., GB, BE, NL) exhibit better class performance. The overall accuracy of about 52% is quite satisfactory considering the relatively high number of classes (the random classifier would have accuracy <10%).

In addition, Fig. 2 illustrates the classifier confusion based on the normalized data. We can see that CZ is the worst predicted class. The classifier has good discriminatory property when predicting users going to FR, CH, IE and IT. In the case of GB, even though it is characterized by the highest true positive rate, its visits are often misclassified to other countries. The same pattern can also been seen in DE and ES. Overall, the confusion matrix shows that FR, CZ, IE, IT and BE are the easiest to be predicted when compared to other countries. we plan to focus on their mitigation in our future work.

4 Conclusions

One of the motivation of this paper is to show how language plays a major role in shaping user mobility behavior, especially, at a country level, as almost every country can be defined by their official language e.g. Spanish as Spain. Hence, in this paper, we have proposed a classification model to predict which country the Twitter user will go based on the user's tweet metadata. From the experimental result, we can see that language does affect user mobility behavior. However, at current progress, only a little improvement is observed when adding language features into the model.

In the future, we would like to improve the classification performance on the whole dataset as well as to include text analysis as part of the classification model e.g. sentiment analysis.

Acknowledgments. This work was partially supported by MIC SCOPE (171507010), and JSPS KAKENHI Grant Numbers 15K00162, 16H01722, 17K12686, 17H01822.

References

1. Baraglia, R., Muntean, C.I., Nardini, F.M., Silvestri, F.: LearNext: learning to predict tourists movements. In: Proceedings of CIKM 2013, pp. 751–756 (2013)
2. Chauhan, A., Kummamuru, K., Toshniwal, D.: Prediction of places of visit using tweets. Knowl. Inf. Syst. **50**(1), 145–166 (2017)
3. Hall, M., Frank, E., Holmes, G., Pfahringer, B., Reutemann, P., Witten, I.H.: The weka data mining software: an update. ACM SIGKDD Explor. Newsl. **11**(1), 10–18 (2009)
4. Liaw, A., Wiener, M.: Classification and regression by randomforest. R News **2**(3), 18–22 (2002)
5. Luo, F., Cao, G., Mulligan, K., Li, X.: Explore spatiotemporal and demographic characteristics of human mobility via twitter: a case study of chicago. Appl. Geogr. **70**, 11–25 (2016)
6. Starnini, M., Baronchelli, A., Pastor-Satorras, R.: Model reproduces individual, group and collective dynamics of human contact networks. Soc. Netw. **47**, 130–137 (2016)
7. Zhao, S., King, I., Lyu, M.R.: A survey of point-of-interest recommendation in location-based social networks. CoRR abs/1607.00647 (2016)

Author Index

Bai, Ting   77

Chang, Yung-Chun   3
Chen, Chong   113
Chen, Hsin-Hsi   196
Chen, Jia   45
Chen, Lijuan   90
Cheng, Xueqi   90

Dai, Hong-Jie   3

Fang, Hui   29, 143

Gu, Qi   77
Guo, Jiafeng   90

Hayashi, Yoshihiko   157
Hirakawa, Koji   157
Hsu, Wen-Lian   3
Huang, Hen-Hsen   196

Inoue, Masashi   126, 164
Inui, Kentaro   67
Ishizuka, Kenkichi   36
Iwakura, Tomoya   67

Jatowt, Adam   203

Kawai, Yukiko   203
Keyaki, Atsushi   60
Kikuchi, Kotaro   157
Kita, Kenji   103
Kobayashi, Tetsunori   157

Lai, Sunny   133
Lam, Wai   133
Lan, Yanyan   90
Leung, Kwong Sak   133
Li, Peng   16
Li, Xiangsheng   173
Li, Xin   173
Liu, Qi   16
Liu, Yiqun   45, 113, 173
Luo, Cheng   45, 173

Ma, Shaoping   45, 113, 173
Makino, Takuya   67
Mao, Jiaxin   45
Matsumoto, Kazuyuki   103
Miyazaki, Jun   60
Murakami, Harumi   189

Nie, Jian-Yun   173
Noro, Tomoya   67

Pang, Liang   90
Pozi, Muhammad Syafiq Mohd   203

Qi, Baoyuan   16

Sekine, Satoshi   67
Shimokura, Masayuki   189
Siriaraya, Panote   203

Ueki, Kazuya   157
Ueno, Hiroshi   126

Wang, Bin   16
Wang, Dong   16
Wang, Yuanyuan   203
Wang, Yue   143
Warikoo, Neha   3
Wen, Ji-Rong   77
Wong, Wai Chung   133

Xu, Hao   29
Xu, Jun   90

Yasuhara, Ryu   164
Yen, An-Zi   196
Yoshida, Minoru   103
Yoshikawa, Hiyori   67

Zhang, Min   45, 113, 173
Zhao, Wayne Xin   77
Zhou, Meilin   16

Printed in the United States
By Bookmasters